An Indigenous Science Education Model

by
Gregory A. Cajete, Ph.D.

This work is dedicated to the "sparkle" in my son James' eyes. It is that "sparkle" which represents the special potential of all Native American and Indigenous youth. It is their sparkle that lights the way to a brighter future for Indigenous people everywhere.

Cover Painting of *The Journeyer* Copyright ©1999 Ivan Eagletail

The cover illustration represents the cultural journey Indigenous People are still making from an ecologically central existence to contemporary society where technology and science modalities prevail. The challenge is to make this journey as a holistic person moving toward a new dawning, sensitive to and fashioned by cultural tradition as well as receptive to modern technology.

Cover Photo: CORBIS/Chris Daniels ©
Cover, Book Design and Graphics: Julie Gray

Copyright © 1999 by Gregory A. Cajete

All rights reserved. No part of this work may be reproduced of transmitted in any form or by any electronic or mechanical means, including information storage and retrieval systems, except for brief reviews, without permission in writing or email from the publisher.

Kivaki Press
gray3@juno.com

Library of Congress Catalog Card Number: 99-73307
ISBN 1-882308-66-2
Cajete; Gregory A.
    Igniting the Sparkle: An Indigenous Science Education Model

Includes Bibliographical References.

1. Indians of North America---Indigenous People---Education---Philosophy. 2. Human Ecology---Worldwide.

Second Edition
First Printing, 1999
Printed in the United States of America

15 14 13 -------- 4 3 2

TABLE OF CONTENTS...

Preface..8

Part One....In Beauty It Begins
Chapter 1   An Introduction..11
Chapter 2   The Students..18
Chapter 3   Native American Education and the U.S. Government..24

Part Two....Science Education and the Cultural Perspective
Chapter 4   The "Cultural/Cognitive Deficit"..........................33
Chapter 5   The Western Scientific Paradigm..........................37
Chapter 6   Redefining Science Education for Native Americans...43

Part Three....Looking at the Learning of Science
Chapter 7   Traditional Native American Education..................53
Chapter 8   Science Learning and the Creative Process.............61
Chapter 9   Ethnoscience, A Native American Perspective......73
Chapter 10 Indigenous Science: An Overview........................81

Part Four....Science as Cross-Cultural Discipline
Chapter 11 The Native American Learner...............................88
Chapter 12 Border Crossings...96

Part Five....The Creative Process in Science and the Native American Learner
Chapter 13 The Curriculum Design as Metaphor...................102
Chapter 14 The Courses..116
Chapter 15 The Power of Myth and Story............................128

Part Six....The Learner Within Bicultural Education
Chapter 16 Teaching Native American Students...................140
Chapter 17 Wholism and the General Systems Theory..........154
Chapter 18 Systems Theory as Mirrored in Native American Ritual and Myth.........................158
Chapter 19 Teaching Strategies...162

Part Seven....Building Bridges of Understanding 182

Acknowledgments..189

Appendix A....Areas for Further Research and Development 191

Appendix B....Science from a Native American Perspective
              Curriculum Syllabi...193
- Syllabus #1   The Creative Process in Art, Science and Native American Cultures..195
- Syllabus #2   Philosophy: A Native American Perspective..199
- Syllabus #3   Social Ecology: A Native American Perspective...203
- Syllabus #4   Herbs, Health and Wholeness: A Native American Perspective...207
- Syllabus #5   Animals in Native American Myth and Reality..211
- Syllabus #6   The Primal Elements: A Native American Perspective...214
- Syllabus #7   Astronomy: A Native American Perspective....218

Comprehensive Bibliography..223

ILLUSTRATIONS

Chart Overview - The Research Design..22

Components/Foundations..23

Tracking a Symbol in the Curriculum Process...........................111

Curriculum Diagrams for the Left and Right Brain..................115

A Curriculum Mandala..118

The Philosophical Aspects of Cultural Difference....................141

Creative Process Instructional Model..171

Lands' Transformational Creative Learning Model..................172

PREFACE...

There is great need for new, creative approaches for teaching Native American and other Indigenous students both the process and the content of science. Because culture shapes the inception and the reception of science, any new approach must include culturally relevant models of instruction and appropriate accompanying materials.

Historically, Native American students have had great difficulty completing their "education" in contemporary Euro-American classroom settings. Based on a longitudinal study of high school sophomores, the U.S. Department of Education (DOE) registers the high school dropout rate for Native American students at 50%, and on some reservations that figure reaches an astonishing 70%! A recent American Council on Education report states that Native Americans account for less than 1% of all college students, and more than 53% of these students drop out after their first year in post secondary education. Why? Is it because these students are less capable? Less inventive? Have less educational fortitude? No. It is primarily because North American educational systems are not structured to be compatible with Native Americans' cultural heritage.

The fact that the special needs, learning orientations and cultural perspectives of Native Americans have been poorly addressed in the teaching of science at all levels of American education amplifies the urgency for developing culturally based science curricula. This sad state of affairs, along with other historic, political and sociological trends in education, has led many Native American students to become alienated from science. That these students score in the lowest percentile in science and math portions of the S.A.T. test and seldom choose to work in science related fields illustrates this alienation.

Yet Indigenous people have applied sophisticated science thought processes for thousands of years. In this book the interface between culture, learning and communication is explored in relationship to the teaching of science. The creative process in science and its integrated reflection in the socio-cultural, psychological, mythological and physical aspects of Native American cultures forms the foundation for a proposed curriculum model. From these aspects are derived the general rationale, course content and strategies of presentation for a culturally and ecologically based science education curriculum model.

The philosophical orientation upon which this work is based is explicated in my first book, **Look To The Mountain: An Ecology of Indigenous Education,** available from the same publisher. In this earlier work I endeavored to illuminate an "ecology" of indigenous education through an exploration of indigenous knowledge bases: the environmental; the mythic; the visionary/artistic; and the affective/communal foundations of tribal life. **Look To The Mountain** brings to the surface the quintessential question, What is education for? This question cannot be answered without deep cultural understanding of the modern crisis of education and the importance of finding face (true character), finding heart (true desire) and finding foundation (true vocation)...all in the context of proper relationship to self, community and the natural world as one strives to become a **complete** man or woman. This is the essence of holistic learning and the foundation of Indigenous Education.

This current work describes a culturally responsive science curriculum that I have been teaching at the Institute of Indian Arts in Santa Fe, New Mexico for twenty-five years, which integrates Native American traditional values, teaching principles and concepts of nature with those of modern Western science. The original research for this book culminated in 1986 in the writing of my doctoral dissertation, "Science A Native American Perspective: A Culturally Based Science Education Curriculum" under the auspices of the New Philosophy Program of International College University Without Walls, Los Angeles, CA.[1]

The primary intent of the curriculum model is to address the alienation many young Native Americans and other Indigenous students experience from science and to redress its lack of relevance. This model, an adaptation and extension of a creative curriculum process research model developed by Robert Zais, also aims to motivate more Native American and Indigenous students to consider science-related fields as career possibilities. Another intent is to enhance the understanding of how Native Americans and other Indigenous peoples historically applied the creative science thought process within cultural contexts. Finally, the model presents an approach for synthesizing the appropriate research on Native American and Indigenous cultural sciences into a form and content that is readily usable by teachers of Native American and Indigenous students.

While the major portion of research and examples used refers to the tribes of the continental U.S., the use of the term "**Native American**" is inclusive of the Alaska Native and Native Hawaiian. **Also, it is important to note that the key concepts, research, methods and model presented in this "case study" are meant to be readily "adapted" to the cultural orientations of other Indigenous populations around the world (hence the word "Indigenous" is capitalized throughout this work) . Given this intention, this work is meant to function as a template and provide insight into the development, transfer and application of cultural knowledge in educational planning in third-world nations as well as among Native Americans.**

Finally, this book adds to the sparse literature on culturally based curriculum and learning science. It will address the relationship between Native American cultural configurations and the way modern American science curricula is perceived, and that will provide teachers with relevant and useful information concerning the socio-cultural dimensions of Native American learning orientations. This work provides a **model** for facilitating interpretation of the enormous amount of research into forms of content and presentation that will enhance the teaching and science literacy of Native American and other Indigenous students.

....

Preface Footnote:
1. Copies of the complete dissertation may be obtained for a zeroxing fee by writing: Gregory A. Cajete, P.O. Box 1167, Espanola, New Mexico 87532.
2. For further reference concerning curriculum development and research, see Zais, R. (1976). *Curriculum Principles and Foundations.* NY: Harper & Row and Knowles, M. (1977). *Modern Practice of Adult Education.* NY: Association Press.

PART ONE.....
In Beauty It Begins

Chapter One......An Introduction

I am an educator of Native American people. What I have been doing and where provides context for understanding what is meant by Indigenous science, and the role I play as a Native American educator. I am a Tewa Indian from Santa Clara Pueblo which is one of six Tewa speaking villages north of Santa Fe, New Mexico. Each of these Pueblos is autonomous but is related to others through custom and language. I grew up on the Santa Clara Pueblo Reservation in New Mexico and was raised partly by my grandmother because my mother worked in Los Alamos, New Mexico where she had to spend long periods of time. As a result, I grew up with my grandmother in a distinctly traditional way. Of course, at the time I didn't know that this was education, as it involved an old style of teaching and old ways of learning, ways children had learned essentials important to my people for hundreds of years. The public school was close by so I didn't have to be enrolled in boarding school, as my mother and grandmother had. I was able to grow up in my own community, which allowed me to also gain a sensitivity to the differences between the way we understood ourselves as Pueblo people and the way the Pueblo community was different than mainstream society.

New Mexico didn't become part of the United States until 1847. It really was its own entity and evolved its own individual identity. New Mexico is very different from any other part of the United States and remains so even to this day. When you grow up in a community with other people of your culture you don't realize your difference. I didn't really understand cultural difference until I was faced with having to adapt to mainstream American culture during my school years. It was then that I realized how different Indian people were, and how we viewed life and education in some very distinct ways. When I was ready to go to college, I had to fight to stay in New Mexico. At that time there was a concerted attempt to recruit Native people from reservations and take them off to Ivy League schools. I was courted by schools like Harvard, Dartmouth and Stanford. Many of my friends did go to those places, but I stayed home. I went to a college that was not far from my home, and that allowed me to maintain constant contact with my community.

After I graduated, I was given the opportunity to teach at a school which had opened in 1962, the Institute of American Indian Arts (IAIA) in Santa Fe. The purpose of the IAIA was to evolve a context in which the artistic potentials of young Native American people from all over the United States could be cultivated and expressed. The IAIA was an experiment in cultural education, an experiment using the arts as a primary vehicle, but also aimed at helping native young people learn about themselves, their culture and their identity. After its opening in 1962, the Institute became famous as a model school when it was recognized by UNESCO as being one of four culturally based schools of note in the world.

For five to eight years it was indeed a shining light in the world of Indian education. But, as is often the case when an entity is connected with the US government, and especially with the Bureau of Indian Affairs, it was vulnerable to the winds of political change. In the 1970's the Institute fell on difficult times and was moved from its original campus to The College of Santa Fe, where it became a "tenant" of that college.

In 1988, Congress enacted new legislation entitled the American Indian Arts and Development Act, which chartered the IAIA as a public/private entity with its own direct congressional funding. It remains a kind of experiment because even this new legislation is designed to see whether this arrangement will work and maintain both its ideals and the congressional mandate.

I began to teach high school science at the Institute of American Indian Arts in 1974. At that time the school had a junior and senior high program as a feeder program for the two year Associate of Fine Arts degree program in the college. During my first year of teaching, I realized that many of the ways of teaching and approaching science, or so called text book science, were just not appropriate for my students. These Native American students came from all over the United States, from urban environments, rural environments. Some were very traditional in terms of their upbringing, others were not. All had a common thread and that was an interest and a willingness to explore the arts. They also possessed a common alienation from science educational approaches they had experienced in reservation and community schools. Charged with making a program work for these students, I put aside all the textbook methods I had brought with me from the Teacher Education College and created new curricula based on my own experiences as a native person. It was a grand experience in that I was allowed to do things that would not have been allowed in another school, certainly not in any public school. I explored and created with my students a process that allowed them to learn in ways <u>they</u> felt good about.

A curriculum evolved over the years. It began with the introduction of native uses of plants in a health science class I was teaching, and it grew into a full culturally based science program. Its story is a story of creation, of the process of interaction in science, art and culture and the integration of those aspects into the expression of a curriculum—a learning, teaching process that actually works well for Native American students who wish to understand and learn about their lost heritage as it relates to science. The curriculum evolved around the idea that every Indigenous culture has an orientation to learning, and that orientation is metaphorically represented in its art forms, its way of community, its language, and its way of understanding itself in relationship to the natural environment that contexts or cradles it. [1]

The problem solution which this book addresses requires an extension of the cumulative influences of ethnoscience, cultural education and the creative process. The insights gained from research in these areas and their implications for the way science is communicated to Native American students form the orienting basis for this work.

Science is a cultural, as well as an individual process of thought and has been utilized in some form by every human cultural group. The processes and products of science and their intimate relationships with human culture form an important part of education.

Research in the area of scientific knowledge transfer in an educational setting based on a cultural perspective is only now being pursued. Much of the work concerns the study of the relationship between scientific and artistic thinking in terms of characteristic brain functioning. The most recent research in this area comes not from science educators but from individuals studying split brain characteristics, cultural learning, creativity, art, cognitive psychology, linguistics, holistic health, theoretical physics and cultural anthropology.

The scope of study in cultural anthropology encompasses literally all human activities — including science. Cultural anthropology is one of the few Western disciplines which attempts to understand a given aspect of a culture as a whole, from the inside, and on its own terms. It is this basic characteristic of the methodology of cultural anthropology which lends itself most readily to the understanding of the realities of Western culture through the "other's" eyes.

Individual attempts to investigate how cultural processes of classification and perception affect scientific thought have been led by anthropologists like Benjamin Whorf and Magorah Marayama. Marayama and others have approached science as a cultural system. Through the examination of the ideas of traditional societies, they have begun to widen the parameters of general scientific thought and knowledge.

One of the major insights into the cultural perception of "separate realities" came as a result of the 1956 field work of Benjamin Whorf among the Hopi Indians of New Mexico. Whorf hypothesized that thought is intimately related to and even guided by a people's language. Implied is the idea that "realities" are different from one culture to the next. In a very real sense we are all wrapped up in our own cultural blanket by our language, worldview and reality, and directly perceive and order the world in reference to this schema. Whorf proposed that Hopi terminology for certain aspects of physical reality reflected a better description of that reality than did modern Western terminology. Western structuring of reality through language does not represent the exclusive legitimate perspective of reality.[2]

Research indicates that there is a "mismatch" between the perspective from which science is traditionally presented in American schools and the general cultural and individual learning styles of Native Americans (McCarthy 1980). After pointing out that "...a system of classification and the conceptual reason for that system as well as behavior in reference to this system forms the essence of 'science' in every culture..." Edward T. Hall, in <u>Beyond Culture,</u> adds "Western science tends to overemphasize the process of classification at the expense of information about the organism...(which) has led Western thought to be predominately preoccupied with specifics to the exclusions of contexting within wholes."

Hall asks: "How can integrative systems of thought be developed from a classification system that fragments and never gets around to putting things together in wholes?"[3]

The research of Richard Blakeslee and Bernice McCarthy exemplifies the insights for applied psychology in an educational context. In <u>The Right Brain</u>, Blakeslee summarizes the characteristic thought processes associated with the right hemisphere of the brain processes (Blakeslee 1980). Bernice McCarthy's highly innovative "4-Mat Model" allows for more holistic and integrative approaches to teaching of learning styles with right-left mode techniques. It addresses the innate individual differences in learning styles which directly affect knowledge transfer (McCarthy 1980).

I would go one step further than McCarthy in pointing out that not only are there differences in learning styles of individuals which reflect right/left brain processes, but there are socio-cultural differences as well. Science in most American schools is presented from a perspective that is heavily oriented toward what McCarthy labels the "type two left-brain learner," a learner who is highly analytical, objective, verbal, structured, and parts oriented. I have observed that many Native American students tend to be intuitive, subjective, non-verbal, synthesizing and oriented to wholes and practical in their application of learning. These characteristics are more characteristic of the other three types of learning orientation identified by McCarthy.

In January, 1975, the American Association for the Advancement of Science Board led by Margaret Mead, passed a resolution that formally recognized the contributions made by Native Americans to the various fields of science, engineering and medicine, and supported natural and social science programs in which traditional Native American approaches and contributions to science, engineering and medicine were the subject of serious study and research.

Based on this mandate, Dr. Rayna Greene, former director of the project on Native Americans in Science for the AAAS, advocated research and development of culturally-based science. Through various studies, insights into the unique problems and perceptions of culturally-based science programs have emerged. Dr. Greene summarized:

"... the lack of Indian participation in science is as much due to an alienation from the traditions of Western science as from a lack of access to science education, bad training in science, or any other reasons conventionally given for minority exclusion from scientific professionalism. Contrary to the general insistence of Western scientists that science is not culture bound and that it produces good, many native people feel that science and scientists are thoroughly Western, rather than universal, and that science is negative." (Greene 1981: 8).

> A difference in perception exists in a science that is directly related to the cultural nature of the society from which it originates. This must be seriously addressed if Native Americans are to increase their active participation in the field of modern science.

The study of the ethnoscience of the Indians of North America is a valuable tool for understanding the cultural influences in science and a way for Indians and non-Indians to gain valuable insights about themselves and the unconscious cultural conditioning of their perspectives of natural reality.[4]

The ethnoscience of each tribe or cultural region is unique and characteristic of that group or geological area in that it reflects adaptation to a certain place. However, "strands of connectedness" and similar patterns of cultural thought begin in the northern polar regions of North America and extend all the way to the tip of South America. The mythical paradigms of the Trickster, the Sacred Twins, the Earth Mother, the Corn Mothers, the Thunderbirds, the Great Serpents, the Culture Hero, Grandmother Spider Woman, the Tree of Life — all exemplify the interrelatedness of Indian cultures. All are extensions of the process of "science" in that they reflect a cultural interpretation based on observation of phenomena and processes inherent in nature. They represent a very primal and artistically metaphoric way of perceiving —a distinctly Native American way of viewing the world.

Until relatively recently, the arts, the hard sciences and the social sciences were presented as totally discrete entities in most American school curricula. Indeed, in many American schools they still are. Such an approach has tended to fragment the human cultural systems being examined, thus perpetuating a distorted perspective of the arts, the sciences and culture in the minds of many students. This approach has been particularly unfortunate for Native Americans.

The fact that science is presented entirely from the Western cultural perspective in most American schools can create a very real psychological conflict for a student raised in a different cultural tradition. It is this conflict and resulting alienation that forms the basic impetus for this work.

As is true with all primal cultures, science as a process of perceiving was and is completely integrated with all aspects of Native American cultural systems. The process of teaching and learning science today becomes a matter of discovering the products and determining how and why these early thought processes evolved into these paradigms within the context of each tribal culture and environment.

When one begins to interpret and translate the symbolic language, art, dance, music, ritual and other cultural wrappings through which these paradigms have been transmitted, one realizes that they reflect perceptive and sophisticated ideas about the essence of nature and the universe. Research by scientists Fritjof Capra, David Bohm, David Peat and others into underlying concepts of many ancient philosophies reveals that many primal sciences have incorporated understandings into their systems that are only now being explored by advanced research in Quantum Physics.

Preliminary attempts are now underway to explore the philosophical foundations and ritual practices of primal cultures using the perspectives gained from ecology, the creative process, split brain research, linguistics, theoretical physics, anthropology and Jungian and archetypal psychology. This re-examination has great potential in that it presents a method of transformation and interpretation of these important paradigms of Native America in the context of the 21st century.

PART ONE...

Chapter Two......The Students

In 1974 when this writer began teaching general biology at the Institute of American Indian Arts High School (IAIA) in Santa Fe, New Mexico, it became immediately apparent that the introduction of Native American cultural content into the presentation of the natural sciences was not only appropriate but essential given the characteristics of the students attending the Institute.

The students at IAIA come from diverse social and cultural backgrounds and represented almost all tribal groups in the U.S. All are of Native American ancestry and share a common interest in art. Their involvement with their Native American identity influences the values, attitudes and behavior of formal education at the Institute.

The Institute began in 1962 as a culturally responsive innovation engaged in a unique educational experiment. The school has served well over 8,000 students through its various programs, including Native Americans from over 100 recognized tribes in North America. Art education which relied heavily on cultural content was an integral part of its mission, philosophy and approach to education. Therefore, the introduction of cultural content in the presentation of science was a natural step.

Despite the Institute's innate receptivity, all of the prior science curricula were based upon nationally standardized science curricula. What Native American content was used had been minimal and poorly organized. I developed and implemented a general, internally consistent, integration of Native American ethnoscience with natural science content.

During the evolution and full implementation of the program from 1974 to 1988: 1) students more successfully retained those science concepts which were integrated with Native American cultural examples as measured by teacher prepared tests; 2) were more highly motivated to learn about Native American cultural sciences in relationship to modern sciences, as measured by student responses and level of class participation and activity; 3) became more positive about science and other academic areas, as measured by student evaluations; 4) retained more through the use of relevant art activities and actual science experiences; and 5) improved their learning and retention with the incorporation of combinations of familiar concrete and symbolic modalities with activities involving kinesthetic, spatial, tactile, visual or musical perceptions.

The students who were the primary focus of this study are Native Americans at the secondary and junior college levels. The targeted group encompasses a broad cross-section of socio-economic, language and cultural backgrounds. Their relative level of enculturation, acculturation and assimilation is equally diverse, as are factors such as age, urbanization, rural orientation and social status. Therefore, this study focuses upon a generalized view reflecting observable patterns rather than specifics.[1]

Over the first five years of the curriculum's implementation, I observed several characteristic patterns of learning styles and communicative orientations. These students:

    1) Tend to learn best that which is related to a familiar set of descriptive relationships.
    2) Prefer to learn in high-context interactional working situations;
    3) Tend to best learn those principles demonstrated by immediate, relevant and practical applications.
    4) Prefer loosely structured and informal settings for learning. A distinct preference for humanistically formatted learning.
    5) Exhibit an overall whole-brain orientation in the processing of information.
    6) Exhibit a consistently creative orientation to problems, though many are predominately oriented to concrete operations.
    7) Exhibit high visual, spatial and kinesthetic orientations.
    8) Have an oral as opposed to a written language orientation.
    9) Think in images rather than words.
    10) Are intuitive, subjective and synthesizing.
    11) Express convergent, divergent or accommodation learning in accordance with the requirements of the situation.
    12) Express their individuality in tandem with their respective Native American cultural identities.

Students who can be called "**rural traditional**" are the least assimilated and have a strong orientation to traditional Native American cultural patterns and personality configurations. They usually live in their tribal communities. Income is derived from available employment on reservations or in communities close to the reservations. Their social and community orientations can be said to be nominally syncretized or interwoven with non-Native perspectives. The preservation of a Native American tribal identity is a visible focus in such groups. For example, many Pueblo and Navajo male students within this group wear their hair tied in a bun with a traditional woven hair tie called a "cango" (a Spanish slang word). Male and female students tended to converse with each other fluently in either Navajo or a Pueblo dialect. The writer found that they viewed themselves as Navajo or Pueblo above any other cultural designations. Most expressed the desire to return and live on their reservations once they completed their education.

Other students can be called "**transitional**" and are characterized by movement toward the assimilation of many American socio-cultural, economic and personality norms in preference to traditional Native American cultural patterns. Members are from both rural or urban areas. Their income is derived from employment sources either on or off the reservation. Syncretization of Native American and non-Native American cultural patterns is readily apparent in this group, however, these students maintain close ties to their tribal culture and community, frequently visiting relatives and friends. Students from this group expressed the desire to gain the best from both their traditional Indian groups and mainstream American society. While they realized that there were many differences between these two cultural orientations, they expressed optimism about being able to achieve a balance between two worlds. These students can understand and speak some tribal language and have basic knowledge of their tribal culture.

The third group of students can be called the "**urban assimilated**." This group is characterized by an almost complete assimilation of American cultural norms. Members of this group are usually of urban orientation, however, a number also come from small town or rural social environments. Their income is derived from a variety of sources within the urban centers in which they live and/or work. Many members of this group are second or third generation "urban Indians" and have only a nominal relationship to their ancestral tribal community. The syncretization of Native American and non-Native cultural patterns is slanted toward American urban cultural norms. Yet "urban Indians" early took the lead in advocating culturally-relevant education to "revitalize" their Native American tribal identities. Native American advocacy and the "banding" together of Native Americans irrespective of tribal affiliations within urban centers led to "Pan Indianism."

Students observed at IAIA reflected co-figurative combinations of these three groups. The majority emulated "American youth culture" patterns, such as enjoyment of "punk rock" or "rap music", hair styles and dress, the result of internalization of mass media and informal socialization within school, family and community. Many expressed poor knowledge of tribal culture and language. However, most of the students from this group expressed a great eagerness to learn about their tribal culture and those of other Indians. Many students were of mixed-blood and expressed concerns about being "accepted." Students from this group tended to be more cosmopolitan in their manners, dress and outlook than students from the other two groups. They dressed in the latest fashion, and were the most articulate in their self-expression. They tended to be "trend setters" both within the social group of the school and in artistic expression.

The data resulting from the project can be considered multi-dimensional and diverse to say the least. Screening was based upon the writer's interpretive application of micro-ethnography, current theoretical positions in bicultural education and personal experience.

The data, and these observations and professional experiences were analyzed with respect to their correlations within the following frameworks:
1) Teaching as a contextual communicative art,
2) Science and Education as cultural systems
3) Creativity in Science and Art,
4) Biculturalism, and
5) Culturally-based curriculum modeling.
The unique advantage of this approach lies in its examination of communication, rules and structure in the teaching and learning processes.

The curriculum model presented in this work evolved from the application of an adapted version of a curriculum research model developed by Robert Zais (1976) and presented in his book, "Curriculum Principles and Foundations." The model which I present is a creative synthesis which can readily be translated into a practical recipe for the planning, development and implementation of a culturally-based curriculum which presents science in a culturally-relevant form to Native American students.[2] (See Diagram on page 22).

CHART OVERVIEW
--THE RESEARCH DESIGN--

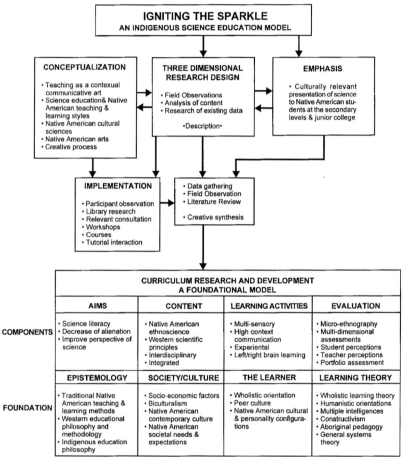

Adapted from Zais, Robert. *Curriculum Principles & Foundation* (1976)

COMPONENTS	FOUNDATIONS
AIMS, GOALS AND OBJECTIVES • To develop a culturally sensitive curriculum model for the presentation of general Science from a Native American perspective. • To address the problem of alienation and lack of relevance many young Native American students feel in reference to Science. • To provide strategies for wholistic learning and a more wholistic utilization of Native American cultural education resources and content. • To address the lack of motivation of many Native American students to pursue Science related professions. • To develop general methods and strategies for integrating the presentation of Science with other disciplines such as Art. • To explore Science from the perspective of creativity and culture.	**EPISTEMOLOGY** • Investigation of the nature and expression of basic Western scientific principles and knowledge. • Exploration of the ways in which Native American cultures symbolically or concretely represented their understanding of nature in their arts, oral traditions, ecological practices, medicine, social organization, architecture and philosophy. • Review of current theoretical literature on culture, cognition, and epistemological of learning inherent in Native American Cultures. (Indigenous education philosophy)
CONTENT/COURSES • The <u>Creative Process</u> in Science, Art and Native American cultural contexts. • <u>Philosophy</u> form a Native American perspective and a survey of Western and Oriental philosophies. • <u>Social Ecology</u> from a Native American perspective and survey of Western social and psychological perspectives. • <u>Herbs, Health and Wholeness</u> from a Native American perspective and a survey of the Western Medical Model, Holistic Health, Chinese and Aryurvedic Medicine. • <u>Animals in Indian Myth and Reality</u> a Native American perspective of animal life and a basic survey of Animal Ecology and Behavior. • The <u>Archetypal Elements</u> an exploration of the Native American Ecological Ethic and Geo-Science. • <u>Native American Astronomy</u> and a basic survey of modern Astronomy.	**SOCIETY/CULTURE** • Relevant characteristics of Modern American post industrial/society. • Characteristics of contemporary American culture and its influence on young Native Americans and their educational needs. • The influence of Traditional Native American cultural values and life styles on attitudes toward science and technology. • The varying effects of assimilation of American culture by individual Native Americans.
LEARNING ACTIVITIES • Exploration of the Artistic and Scientific creative process through hands on art and science activities. • Experiential/Exploratory learning activities. • Right/Left Brain learning activities. • Creative Play, Storytelling, Role-Playing, Creative Writing and Simulation. • Field trips, Guest speakers, Film and Workshops. • Lecture/Demonstration • Individual and Collective Research • Tutorial activities. • Laboratory activities.	**THE LEARNER** • A general cultural Wholistic learning orientation. • Creatively inclined and oriented toward concrete operations. • Exhibits high visual, spacial or kinesthetic orientations. • Oral language orientation vs. Written language orientation. • A convergent, divergent or accommodation oriented learner. • Intuitive, subjective, non-verbal synthesizing and oriented to wholes. • Artistically inclined. • Image thinking vs. Word thinking.
EVALUATION • Pilot evaluation of the curriculum model by selected schools. • Independent consultant evaluation of the model's effectiveness in addressing the stated purposes of the project. • Administrative evaluation of the model. • Students evaluation of the model. • Teacher evaluation of the model. • Comprehensive evaluation of the model by the principle researcher.	**LEARNING THEORY** • Humanistic/Wholistic learning theory. • Traditional Native American styles of teaching and learning. (Aboriginal pedagogy) • Right/Left brain teaching and learning styles. • High context/context learning environments. • Constructivism. • Gardner's Theory of Multiple Intelligence (M.I.) • General systems theory. • Creative learning methologies.

PART ONE...

Chapter Three......Native American Education and the U.S. Government

In 1974 the General Conference of UNESCO proposed a program of concentrated research "... stressing the preservation of cultural identity, authenticity and dignity possessed by each national group." The conference recommended that transfer of knowledge endeavors be predicated on these realizations: 1) an obvious imbalance between the world's producers of knowledge and its consumers which reflects relationships between developed and developing countries; 2) the transfer of knowledge as a base for political power, as the origins and flow of scientific and technical knowledge directly affect political policies and cultural identities, authenticity and dignity; and 3) preservation or revitalization of "endogenous" approaches.

In short, <u>knowledge must be considered to flow not only from the more developed to the less developed, but from the other direction as well.</u> The nature of education available to a given cultural group, and what is learned through both the formal and informal (hidden) curriculum, communicates how that group is perceived from a socio-cultural and political standpoint. Knowledge is power. Those who have ready use to an education which prepares them for such a society have the greatest opportunity and impetus to develop that education into an economic and political power base.

Science education plays an integral role in the development of such a power base. Yet every culture has its own system of ethnoscience which directly affects the way modern science education is received and perceived. It therefore becomes very important to consider the nature and effects of "endogenous" approaches to science in order to develop a positive syncretization between modern Western science and other cultural sciences (UNESCO 1978).

The UNESCO research program emphasized sharing knowledge as opposed to using knowledge to dominate others which leads "to profound disparities in economic, educational, and social status, a devaluing of cultural heritage and diversity, limitation of cultural creativity, internal friction, increased depersonalization and alienation" (UNESCO 1978).

The UNESCO Division for the Study of Development has concentrated on basic research in this area since 1974. The range has included studies regarding the cultural nature and dimensions of human cognition, power relationships concerning knowledge transfer, the flow of knowledge in cultural contexts, languages and systems of reasoning, intellectual and cultural creativity, and the mediums and levels of knowledge exchange.

At the national level, the American Indian Science and Engineering Association (AISES) has focused upon the specific characteristics and needs of Native American learners and communities in response to the practical need for communication about science education and science employment opportunities. Activities have included the exchange of information between science professionals, industry, local educators, tribal officials, colleges and universities and others concerned about science and mathematics education. In addition, AISES has become involved in the development and implementation of science and math programs which are directly relevant to the cultural identities of Native American students in order to build bridges between science as it is presented in the classroom and cultural perspectives.

Research in planned cultural approaches to teaching and learning is only now moving from its formative stages to new and more creative expressions. The real strength of current research lies in the fact that synchronistic lines of thought and similar conclusions are beginning to appear throughout the literature. This has in turn stimulated the formation of networks of researchers and other interested groups at the local, regional, national and international levels.

The main weakness of current research appears to be the lack of quantitative data concerning how, why, and in what contexts culturally-based approaches are more effective than conventional approaches. Research in this area is only in the first stages of development and quantification will follow as information and current findings become more widely disseminated or combined with other research involving cognition, creativity and brain research. Unfortunately, some researchers are already downplaying the validity of culturally based curricula based solely on the fact that few studies exist to verify their effectiveness.

Yet, such downplaying is understandable since many researchers tend to emphasize one or two dimensions of what in reality is a dynamic, multi-dimensional and process oriented entity. For instance, the current emphasis on science literacy, entrance into science related fields, and relevant content, while defining pressing needs among Native Americans, is only a preliminary step toward addressing greater and more acute educational, social and economic needs. Other dimensions of research of a cultural approach in areas such as creativity, cross-cultural communication, interdisciplinary orientations to arts and sciences, holistic learning theories and health remain in the first stages of development.

Few studies have been concerned with the presentation of science as a cultural system. Rather, most studies have attempted to adapt cultural content to the subject matter as it is contained within the standard school curricula. Few studies, especially by educators, have challenged the cultural premises of Western science and its reflection in the content and presentation of science curricula. Yet such an approach to science education by its nature and process brings the assumptions of Western science to the forefront. These can then be examined and critically compared to the assumptions of other cultural systems.

The idea of a culturally-based approach to science education for Native Americans is a new development in a long and tenuous history of Native American education and reflects an evolution of thought related to self-determination, community education and a renaissance of Native American identity.

Over the course of contact between European and Native American cultures, the sustained effort to "educate" and assimilate Native Americans as a way of dealing with the "Indian problem" inevitably played a key role in how Native Americans have historically responded to American "schooling." As is true with many "colonized" cultural groups throughout the world, the first attempts were met with resistance. This resistance to schooling continues today in varied and psychologically submerged forms.

Historically, the first attempts at introducing Euro-American schooling to Native Americans were met with suspicion, apathy and indignation. However, once education became viewed as an essential aspect of adapting to modern society, it rapidly evolved into an indispensable key to personal and tribal success in direct proportion to the assimilation or adaptation of "core" American cultural values. This general scenario was played and replayed for each Native American tribal culture.

Native American cultures first encountered Euro-American educational forms through contact with missionaries. Missionaries consistently focused upon changing the religion of Native Americans, which in most instances necessarily entailed a drastic change in the tribal culture as a whole. Such schooling directly conflicted with traditional forms of Native American education. Early missionary and government teachers naively assumed that Native Americans had no education at all, and that their mission was to remedy this "great ignorance." The notion that the learning process is adapted to the environment and cultural configurations of that society did not occur to educators until the late 1920's.

In reality, every Native American tribe had its own system of enculturation that involved a wide variety of learning strategies. The available research concerning the educative processes in many "primal cultures" suggests that a child was educated and educated himself through various formal and informal interactions from birth to old age involving both simple and highly complex forms of education (Murayama 1978: 173).

Traditional Native American systems of educating were characterized by observation, participation, assimilation and experiential learning rather than by the low-contexted formal instruction characteristic of Euro-American schooling. Among the high cultures of Mexico, Central and South America, education was highly interactive, formalized by tutor/student relationships and established elite "academies" which creatively integrated formal with informal learning and teaching. Education was highly dependent upon intrapersonal, interpersonal, kinesthetic and spatial learning as expressed in oral language and active involvement within the tribal culture.

Throughout the eighteenth, nineteenth and into the early twentieth centuries official U.S. government policies were geared toward the total assimilation of Native Americans. Toward this end, the government established the first boarding schools in Oklahoma in 1887. Eventually these were established throughout the Western half of the U.S. Early boarding schools were "deculturalization centers," which emphasized the vocational curriculum of "the primer and the hoe" as a replacement for Native American traditions and language (McBeth 1984).

The "New Deal" in Native American education began after the Indian Reorganization Act in 1933. This monumental act was based on the Merriam Report (1928), the first comprehensive study of Native Americans, which outlined their social, economic and educational plight.

Initial progress in education was shadowed by recurrent problems of alienation, bureaucratic chaos, societal barriers and apathy. Influencing these problems were the assimilationist ideologies and ultra-conservative influences within the federal government, which eventually culminated in the termination of responsibilities for several Native American tribes in the 1950's.[1]

Over the years Native Americans have been able to adapt even the boarding school systems to the socio-cultural tapestry of contemporary Native America. In its contemporary expressions the boarding school system may have fostered unintended intercultural communication and other culturally adaptive benefits. This subversion of the original intent of boarding schools from "deculturization" centers to modified "enculturation" centers re-enforcing "Pan Indian" communication illustrates the tenacity and ingenuity of Native Americans. Separation of Native Americans from the dominant society may have inadvertently reinforced students' Native American identity and facilitated the "informal comparing of notes" and the formation of lifelong friendships and understandings. This "hidden" conditioning, as well as the perception of some that contemporary Indian boarding schools are strongholds of contemporary Native American education, underlies much of the current opposition to the closure of these schools (McBeth 1984).[2]

During the 1960s, Native American leaders met in various parts of the country to discuss common problems and a host of social and political issues. As a result, the federal government commissioned a special task force which in 1969 published a report entitled "Indian Education: A National Tragedy, A National Challenge." The report blamed failures on bureaucratic malaise, overwhelming poverty, community disorganization and waste, misinformation, stereotyping, prejudice, discrimination and educational approaches that disregarded Indian identity and learning styles (Snow 1974).

This research revealed highly influential and interrelated factors which helped form the attitudes, behavior and relative achievement of Native

American students. A child experienced a "mismatch" between the educative process and transfer of cultural tribal configurations. Teachers and schools were not geared to Native American mindsets. This led to misunderstandings, prejudice, stereotyping, alienation of Native American communities from schools and school personnel, and teaching methods and content which were out of sync with Native American cultural contexts. The rapid urbanization of Native Americans had and would continue to have definite implications for educational approaches.

The 1967 Havighurst study recommended that Native Americans be given the deciding influence on educational matters which directly affect them. A national advisory board on Indian education should be formed to do further research and to make recommendations for "de-homogenized" (non-standardized) school curriculum content tailored to student needs.[3]

The first two recommendations have been implemented and are beginning to show positive results. However, the "de-homogenization" of curricula offerings remains in the preliminary phases of development.

The Josephy Study of 1968 indicated that woes in Indian education could be traced to the "deaf ear syndrome," or an inability of Westerners to listen to or accept Indian recommendations for change. The study indicated a lack of vision, historical perspective and understanding of the Indian experience (Snow 1974).

Between 1972 to 1978 legislation began to address the findings of the studies. The Indian Education Act of 1972 (PL 92-318), the Indian Self-Determination and Education Assistance Act of 1975 (PL 95-638) and the Education Amendments of 1978 have had a marked impact on the quality of educational services for Native Americans. However, progress has been slow and characterized by substantial bureaucratic and political ineptitude. The progress made by Native American tribes in their efforts in self determination will directly depend upon their success in adapting modern American education to their cultural, economic, and political needs.[4]

As these studies had been highly critical of the Bureau of Indian Affairs, a variety of changes were made during the 1970s within the B.I.A., especially in the Southwest where Indians remained relatively unassimilated.

Ineffective communication networks between B.I.A. educational administrations and educational personnel in the field as well as the tendency of top level administrators to "sit on information" significantly hampered the dissemination and/or implementation of research findings. Despite such communication problems, two studies were completed from 1972 to 1974 which reached similar conclusions concerning culturally-based approaches to educating Native Americans.

The first of these studies provided a substantial focus for a culturally-based approach. In 1972 Ralph L.Casebolt reviewed the enculturation process at Zuni pueblo from historical times to the present, tracing insights which could be applied to curriculum development. If educators understood enculturation, they would come to experience student mindsets and would be enabled in developing more culturally relevant school environments. Cultural anthropology, Casebolt demonstrated, could be used as a tool and a resource.

The second major study completed in 1974 by Albert Snow provided statistical evidence to show that the use of cultural content in science was not only plausible, but that it enhanced the learning of Native American students in a measurable way. Snow felt that unless a relevant socially and culturally meaningful way of presenting modern science to Native Americans was developed, Native American cultures and modern technology would continue to be characterized by a sort of "future shock syndrome." He recommended further research into how to integrate culturally-based science content into sense perception learning experiences, development of content materials for ethnoscience curriculum, the development and implementation of practical teacher training programs especially for non-Indian teachers, and greater involvement of Indian educators and the Indian community in the ethnoscience classification process and in curriculum development.

Through her early work as director of the "Project on Native Americans in Science," for the American Association for the Advancement of Science, Dr. Rayna Greene was instrumental in developing awareness of Native American cultural sciences and in supporting the study of culturally-based science education. Dr. Greene conducted yearly seminars on culturally-based science research and its applications to education, health and economic development. She studied the reasons for the lack of participation of Native Americans in science and found that "the lack of Indian participation in science is as much due to an alienation from the traditions of Western science as from the lack of access to science education."

"Contrary to the general insistence of Western scientists that science is not culture bound and that it produces good," continues Dr. Greene, "many native people feel that scientists are thoroughly Western, rather than 'universal,' and that science is negative. They insist that the practice of Western science trains alien, unfeeling people who bring environmental and human damage in their wake. More importantly, they perceive themselves as among the increasing numbers of endangered species for which they hold scientists responsible. With good evidence on their side, Native people fear Westernization, and the consequent alienation from their communities of tribal members who become 'scientists' in the Western manner. They fear that Western science is an antithesis of traditional religious belief — a fear also expressed by many Christian traditionalists — and that Natives trained in a Western tradition will lose their respect for 'old ways'" (Greene 1981: 8).

This alienation of Native Americans toward science and scientists, combined with the perception of many scientists and educators that Native American cultural sciences are little more than primitive folklore, has led to "closed" conceptions of the cultural nature of "science" on the part of both parties. In turn, this "cultural mis-communication" has directly affected the reception and presentation of science in formal educational contexts.

Dr. Greene views culturally-based science as "humanized."

"Many would also say of the natural sciences that their very obsession with objectivity is the key to their inhumanity, and that a counter-influence, to be found in non-Western, non-rationalist traditions, would multiply the maximum benefits of science ... culturally-based science can both bring a new kind of knowledge — a knowledge which changes the scientific disciplines — and humanizes those disciplines simultaneously" (Greene 1981: 8).

PART ONE...
Footnotes

Chapter 1
1. For a more in depth perspective of the development of an Indigenous teacher see Cajete, Gregory A. (1999). The Making of An Indigenous Teacher. In Kane, Jeffery (Ed). *Education, Information, and Transformation.* Prentice-Hall Pub.
2. See Whorf, Benjamin (1956). *Language, Thought, and Reality.* (215-218). Cambridge: MIT Press.
3. See Hall, Edward T. (1976).*Beyond Culture.* NY: Anchor/Doubleday.
4. For further elaboration of the studies mentioned herein, see Snow (1974).

Chapter 2
1. The characteristics of students outlined heretofore have been derived from application of informal participant observation techniques over the first five years of the curriculum's implementation. To formally verify and extend these observations would necessarily require the application of appropriate evaluation instruments. The writer feels that the application of such a formal evaluation would support the informal observations which have been presented.

Chapter 3
1. For further elaboration concerning the circumstances surrounding Indian education during this period, see Whiteman (1985), and Fuchs (1972).
2. See McBeth (1984) which presents the ways the B.I.A. Boarding School system may have played a vital role in the perpetuation of contemporary "Pan-Indianism."
3. See Introduction, Szasz (1983) in reprint of <u>To Live on This Earth</u> by Estelle Fuchs and Robert J. Havighurst.

PART TWO...
Science Education and the Cultural Perspective

Chapter Four......The "Cultural/Cognitive Deficit"

The average test scores of Native Americans in science and math are significantly lower than those of their white American counterparts. The few studies which have been conducted in this area show that all students significantly improve their performance if they are given the proper guidance and environment. Factors which have been found to affect both guidance and environment include: opportunity for learning, socio-economic status, teacher preparation, self-concept, parental encouragement and assessment of learning potential (Ovando & Collier 1985: 392).

Over the last century an enormous amount of ethnographic material has been accumulated on Native American cultures, including information on traditional teaching and learning methods. Although traditional Native American methods of teaching and learning have been altered considerably through the process of acculturation, the ways in which students learn and the extent to which they learn in culturally-conditioned ways are key elements in making science relevant to them. A synthesis of such accumulated research can provide the needed framework for providing a more relevant science education for Native Americans.

Different groups of people respond to the same environments in different ways. Yet, in the early 1960s much research concerning minority students' problems with math and science revolved around the mistaken concept of the "cultural/cognitive deficit." Research based on the cultural/cognitive deficit involved viewing the ways in which thinking skills and educational attitudes were hampered by the home environment, which necessarily also implied the cultural environment.

Many of these studies inadvertently implied a superiority of mainstream American culture. Some studies graphically revealed a bias in their statements to the effect that minority groups having problems in the areas of science and math had internalized a cultural configuration which needed to be changed if they were to realize success and fully participate within a typical American school context. (Ovando and Collier, 1985.)

It was generally assumed that the home environment was in great need of alteration toward "more American" cultural patterns if such students were to succeed in school. Very little research was attempted in ways to reaffirm the culture which minority learners brought with them and to syncretize learning activities to fit the predominant learning styles. A few researchers like Robert Havighurst argued that it was important to understand how schools themselves were responding to the culture of their students. Havighurst felt that schools were in fact causing many of the problems which minority students were experiencing because they failed to enhance or in any way support the cultural identities of the minority students which they served. (Havighurst 1955).

This early emphasis on the cultural deficit was very much a part of early programs in Native American education. This frame of reference was used by some educators as justification for ignoring the unique cultural knowledge and orientation to learning which Native American students brought with them from home.

In many ways, the cultural/cognitive deficit orientation of that era delayed all educators from realizing the benefits of multicultural perspectives. Echoes of the cultural/cognitive deficit approach continue to be heard from many in the continued trend toward the homogenization of science curricula and in the positions of those educators advocating "back to basics" ideologies. However with the increasing momentum of the technological age and the fact that the "old basics" in science are becoming increasingly obsolete because of the rapid rate of knowledge "turnover," it is clear that the call for "back to basics" must become a call forward to new basics.[1]

These "new basics" must include greater sensitivity to the cultural influences on science learning and recognition of differences as <u>complementary</u> to the learning of science rather than an obstacle to be overcome by science educators.

What may be needed is a theoretical framework from which to view science learning as a thought process. The work of the Soviet psychologist Vygotsky is useful in the exploration of science learning from this perspective. Vygotsky focused primarily upon the way problems are solved and on the cultural context in which the development of learning takes place. His thesis revolves around the relevancy of scientific or technical knowledge to the requirements of the task at hand. Another aspect deals with the culturally-based knowledge internalized by the individual which can be applied to contexts within the individual's reference group. Problem solving takes place as a result of the interplay between these two influences. A cultural context guides approaches which children will eventually spontaneously apply to other problems they encounter (Saxe & Posner in Ovando & Collier 1984: 199).

There are a host of examples within particular cultural contexts where learning of "applied science" takes place. A few in Native American contexts might include: a girl accompanying her grandmother on a herb gathering trip, a boy watching his father hunt deer, a girl watching her mother dye wool, a boy watching his grandfather observe the sun and moon to determine when to plant or harvest corn, or children watching their parents involved in a particular art form. Children internalize a cultural style of thinking and problem solving at a very early age. As they grow older and become more personalized in their problem solving and intellectual competence, much of their thinking still reflects their earlier cultural conditioning (Ovando and Collier 1984: 199-200).

While Vygotsky recognized that children do learn directly through interaction with their environment, he also saw that thinking is always a mediated activity in which the growing child "interacts with representations (symbols) of the world which include culturally rooted systems as language and numeration" (Saxe & Posner in Ovando & Collier 1985: 198).

Research based upon the Vygotskian thesis provided evidence that skills children develop in numeration during the concrete operational stage (7-11 years) are related to culturally-based numeral systems and the practices surrounding their skills. The Japanese, for instance, are skilled in certain types of computation because of their early familiarity with the use of the abacus and because "the Japanese language provides regular rhyming words for the memorization and multiplication of facts." Similar findings relating numerical competence with specific cultural requirements have been reported for the Kpelle of North Central Libera. Competence in a given dimension of logical/mathematical thinking is directly related to what these children are familiar with, have an interest in, and are consistently involved with through their culture (Saxe & Posner in Ovando & Collier 1985: 198-200).

From such research observations, it may be said that science learning should be highly kinesthetic and activity oriented, using a variety of sensory modalities in creative combination. Cross-cultural studies have shown that this is the way both science and art are learned within traditional cultural contexts. Familiarity, using, tasting, feeling, hearing, seeing, smelling, manipulating, and all the combinations therein evoke the learning instinct and have been used by various cultures to teach what was felt to be important to know about the natural world.

Culture and learning are intricately interrelated. Early educators attempted to minimize the recognition of cultural influences other than those of mainstream American society. It is now recognized that the whole of culture greatly influences all levels of learning. The "cognitive/cultural deficit" frame of reference has given way to a more enlightened recognition of cultural orientation as an essential consideration in the learning process.

The implications of an underlying paradigm for science learning are enormous. They must become integral considerations in the development of science education curricula.

PART TWO...

Chapter Five......The Western Scientific Paradigm

What is taught as science in modern schools often leads students to internalize a distorted view. This distortion revolves around the perception that the scientific method, if followed explicitly, is infallible, the completely objective and unbiased way to uncover the facts and reach truth. (Hayward 1984: 66).

Teaching the basic concepts forming the foundations of modern science, students are led to believe that:
• time is uniform and flows in a single linear direction from a past to a present and on to a future;
• matter is made of particles that obey universal laws which never change;
• our mind is our brain;
• only the fittest survive through the process of natural selection;
• modern science will eventually solve all the major mysteries of the universe; and
• **scientists are totally objective and scientific knowledge is universally applicable.** [Hayward 1984: 66].

Students are taught these "realities" because scientists and science have found these things to be factually so, and therefore anything else is mere speculation, myth or fantasy.

Yet, there have always been and probably will always be realities which directly contradict and are anomalous to what modern science contends is so. These realities derive from other cultural systems of scientific thought which have evolved from unique perceptual orientations of natural reality. For instance, among some Native American healing traditions, there is a perception that disease can be caused by a "wayward spirit" of the natural world that enters the body and causes disharmony. Within the structure of this Native American paradigm, this "logical" perception orients Native American medicine in a different direction than Western medicine.[1]

Yet anthropologists insist that no cultural paradigm of reality (including that of Western science) can be based totally upon a completely rational and objective perception. Cultural perceptions, upon which all paradigms are ultimately based, are holistic phenomena and are necessarily influenced by a host of interrelated factors which extend far beyond rationalistic thinking.

Jeremy Hayward argues that:
"The so-called scientific method which, Johnny learns, is how scientists work is a fabrication.... It is extremely unlikely that there can ever be such an event as a pure, unprejudiced observation. Observations are always made, first, within the historical and cultural context which has given someone the reason for making the observation at all and, second, within the context of the person's own belief system and his own organism which in turn affects the result of his observation." (Hayward 1984: 71).

Hayward contends that this often hidden and mostly unconscious characteristic in regard to the Western Scientific Method leads to a conditioning of students and scientists alike which can be roughly translated to mean, "I'll see it when I believe it...This principle appears to be a very deep principle of the organism, a principle which governs much of our perception. Simply, this view proposes that the conventionally agreed upon belief structure of any particular group of scientists forms an inseparable part of their practice of science and determines which observations of science will be accepted by them and which will be rejected. From this point of view, then, science is not necessarily finding out the truth, but merely confirming or refuting its agreed-upon belief structures. In addition, belief systems change only when, in confrontation with reality, error is revealed." (Hayward 1984: 71-73).

Then:
1) The way in which each of us perceives and understands the world is directly dependent on the unique configuration of our belief systems;
2) The meanings which we attach to elements of natural phenomena are directly dependent on the conceptual structure of which they are a part, highly conditioned by the culture and system of thought; and
3) What constitutes "fact" is directly dependent upon the consensus of the community or group which evaluates what is real and what is not based on a mutually held belief system. (Hayward 1984: 76-77).

The scientist is viewed as an impartial and objective spectator of the phenomena which he studies. The processes which he observes and makes notes on and later analyzes to derive generalizations are there for anyone to see. Generalizations derived from scientific observation are called "hypotheses." This perception of scientific work is in reality simplistic. Observation and hypothesis formation always lead to more observation and hypothesis formation in a cyclic process of dynamic and creative response to what is being studied.

Ideally, the dynamic process of observation and hypothesis formation is meant to be totally objective. That is, "facts" are in the world to be seen by anyone looking for them. Theories are systematically and logically developed around the facts. Complementary or contradictory theories are judged on their merits and relative compliance with the scientific method. Whether a theory is judged valid is based not on reasons that are subjective, political or cultural but rather reasons that are objective, rational and open to public scrutiny. Theories are validated through a consensus of the scientific community, which compares the theory against the body of knowledge currently known. Yet this ideal is not the reality of the application of the scientific method because of the conditioning of the underlying structuring paradigm.

To understand a cultural system of science, one must first "get inside" the paradigm of the cultural community within which that system operates. Thomas Kuhn, in a history of modern science reprinted in his work in 1977, explores the cultural paradigm of Western science. Kuhn defines the word "paradigm" as the basic grammatical practice of memorizing declensions of nouns and conjugations of verbs in the study of languages like Greek and Latin. He first uses the word in his 1959 essay, "The Essential Tension." There he extends its meaning: "scientific paradigms" are exemplary problems such as the inclined plane or the pendulum, the study of which initiates the student into the scientific community.

But paradigms, Kuhn points out "took on a life of their own."
"They expanded their empire to include, first, the classic books in which these accepted examples initially appeared and, finally, the entire global set of commitments shared by members of a particular scientific community." (Kuhn 1959: 19).[2]

Paradigms, or the global set of shared commitments, divide as well as unite scientists. In "normal science," research is carried on within the boundaries of shared beliefs. But in "extraordinary" or "revolutionary" science, contradiction is discovered and is followed by the upheaval required to get rid of one paradigm and acquire another (Austin 1982).

Normal science is largely a mopping-up job, based in scientific achievement — say, the publication of Newton's Principia. Such an achievement implicitly defines the methods of the science community and the sort of data worthy of consideration. Many problems will be excluded; they will be considered either already solved or not worth the effort of attempted solution. But a few problems will remain. These problems will take on the aspect of puzzles for the scientists to crack open. **Failure here will reflect more on the scientist than on the problem** (Austin 1982).

For instance, Newton's Principia Mathematica (1687) established the idea of universal gravitation. After it was accepted as an achievement forming the basis of a paradigm or shared belief, no scientist had to worry about whether or not gravity was indeed an inherent quality of all matter; that question could be excluded from serious consideration. Instead, scientists concentrated on ironing out the details of universal gravitation. Newton had developed the theory to the point where, knowing how strong was the attraction between the sun and any single planet, he could derive that planet's orbit. These ideal orbits are only approximated, however, by the planets' real orbits. The planets exert gravitational pull on each other, a factor that had to be worked into the theory: giving mathematical form to the deviations allowed the observed orbits and the theory to fit together better. Some of the greatest mathematicians of the day worked on this puzzle. It is interesting to note that, in spite of success elsewhere, they could never account for Mercury's orbit because its orbit was an anomaly which didn't fit the paradigm. **Scientists did not conclude that Newton was wrong, but only that they had yet to apply his theory properly to the data** (Austin 1982).

Not only is most "normal science" conducted within an unquestioned paradigm, but normal science is essential to the progress of science. Only by excluding broad questions can scientists concentrate on promising research. The normal scientific procedure is to make every effort to force a recalcitrant fact into the paradigm or to make it go away. For the paradigm is the unquestioned foundation of science as its practitioners know it. Most have never worked under any other paradigm; indeed, most are simply unaware that they even have a paradigm. The failure, then, of a fact to fit a basic theory would reflect not on the theory but on the scientist (Austin 1982).

Kuhn claims that paradigms also determine which facts will be seen. He even claims that in some sense the facts themselves are changed when paradigms change: that after a paradigm shift one is living in a different world. He shows how scientists systematically ignore data which cannot fit into accepted theory, unless that data is forced upon them. He shows that observation is a complex process in which the perception and its interpretation are inextricably bound together. How is it that scientists can proclaim their own objectivity when they are in fact engaged in such a narrow pursuit? The answer is that they must be unaware of their paradigms. How is it that they are unaware of them? The answer to that question lies in the function of science textbooks.

"The single most striking feature" of science education, claims Kuhn in <u>The Essential Tension</u> (1959), is that, to an extent totally unknown in other creative fields, it is conducted entirely through textbooks. Kuhn continues:

"In music, the graphic arts, and literature, the practitioner gains his education by exposure to the works of other artists, principally earlier artists. Textbooks, except compendia of or handbooks to original creations, have only a secondary role. Contrast this situation with that in at least the contemporary natural sciences. In these fields the student relies mainly on textbooks until, in his third or fourth year of graduate work, he begins his own research. Until the very last stages in the education of a scientist, textbooks are systematically substituted for the creative scientific literature that made them possible." (Kuhn 1977: 164).

Most science textbooks contain the facts, theoretical constructs and exemplary problems which form the foundation for the current scientific paradigm. Textbooks, Kuhn writes, "...aim to communicate the vocabulary and syntax of a contemporary scientific language...(They) record the stable outcome of past revolutions and thus display the bases of the current normal scientific tradition. ...they then ask the student, either with a pencil and paper or in the laboratory, to solve for himself problems very closely related in both method and substance to those through which the textbook or the accompanying lecture has led him." [3]

Writers of science texts unknowingly "mythologize" the association between a particular phenomena and the scientist who first discovered it. Kuhn writes:

"Characteristically, textbooks of science contain just a bit of history, either in an introductory chapter or, more often, in scattered references to the great heroes of an earlier age. From such references both students and professionals come to feel like participants in a long-standing historical tradition" (Kuhn 1977: 136-137).

PART TWO...

Chapter Six......Redefining Science Education for Native Americans

As Native American communities take control of more of their own education, the integration of traditional and contemporary values in quality education becomes a greater possibility. Presently, little improvement in science education has been realized. Few schools which serve Native students have integrated cultural content in any serious or systematic form. This lack of progress is reflected in the continued underachievement of Native students in science and math. The need for expertise among Native people in the area of science has never been greater because of scientific and technical literacy and skill needed to effect self-determination in tribal resource management, health and economic development (MacIvor 1996).[1]

Encouragement and support is crucial in the development of a foundation in science literacy. However, given the poor results achieved during the prior history of Western education of Native students, there is a need for a radically different approach. Such an approach necessarily requires the development of a new view of native education in which the new teaching and learning may be contexted. What would a modern indigenous philosophy of education consist of? Curricular change in native education must stem from "cultural standards" firmly grounded in Native thought and orientation (Ibid).

Eber Hampton outlines 12 standards for a Native American education.

1. Spirituality: Respect for spiritual relationships.
2. Service: To serve the community given its needs.
3. Diversity: Respect and honoring of difference.
4. Culture: Culturally-responsive education process.
5. Tradition: A continuance and revitalization of tradition.
6. Respect: Personal respect and respect for others.
7. History: A well developed and researched sense for history.
8. Relentlessness: Honing a sense of tenacity and patience.
9. Vitality: Instilling a vitality in both process and product.

10. Conflict: Being able to deal constructively with conflict.
11. Place: A well developed researched sense for place.
12. Transformation: The transformation of Native education.[2]

These 12 standards as applied to science education for the Native learner can be adapted as follows:

Spirituality is an integrated focus on traditional expression of indigenous science; as such, it should be an equally integrated consideration for the teaching of science to Native students. There are Native expressions of metaphysics which focus upon learning about and then applying embodied relationships with the natural world. The spiritual dimension includes all aspects of human experience and is the ultimate expression of relational understanding in the perspective of traditional Native American societies. The integration of spirituality and learning can be problematic unless a well developed understanding of cultural appropriateness is applied in the presentation of science content.

Service is an integral value of traditional Native education. This is to say that the value of any form of education is how well it can be applied to the needs of the community. From this perspective the value of science and technical education is gauged by the extent to which it may be applied to the solving of community problems. However science learning is largely divorced from community needs. This means that "bridges" must be built to connect the learning and content of school science to the needs of Native people and communities. This can be accomplished by ensuring that science content is tailored to a specific community life and its needs; the application of service learning methodologies and understanding must be a coordinated community effort.

In terms of diversity, a concerted effort must be made to develop a sense of the multi-tribal, multi-cultural perspective in the application of science literacy in an increasingly multicultural world. The role of culture in terms of content and process is integral to all aspects of a Native based science curriculum. The role of language as carrier and process medium is a key consideration since science can be viewed as a form of story and a kind of language. This means that ESL (i.e., English as a Second Language) methodologies have particular applications in science learning for Native American students.

ESL instruction has been succinctly defined as: "a structured language acquisition program designed to teach English to students whose native language is not English" (Briere 1979: 201).[3]

[Bilingual education is]...the use of two languages, one of which is English as mediums of instruction for the same pupil population in a well-organized program, which encompasses all or part of the curriculum and includes the study of history and culture associated with the mother tongue. A complete program develops and maintains the children's self esteem and legitimate pride in both cultures (U.S. Office of Education 1971).[4]

Learning style theories can also have application in science learning but must be used cautiously to avoid categorizing students based on dispositions that "stereotype" students of Native American populations. In researching Native American learning styles, one needs to keep in mind the great diversity between and within tribal groups. For example, research has shown that Native American students have visual-spatial strength, a tendency toward global ability and reflectiveness. However, this tendency is filtered through the relative level of students' acculturation to school culture and expectations, inconsistent or inappropriate testing (More, 1990, 6-9).[5]

There is a traditional history of science application for each of the hundreds of tribal groups of Native America. Native students need to be provided with the opportunity to learn about these histories with specific attention to their tribal affiliations. Attention to the tribal histories of science will give students insights into the wealth of traditional knowledge related to science and in the process give them a foundation for pride in their peoples' accomplishments and a more authentic view of the history of science (MacIvor 1996).

The relationship of respect for oneself and the knowledge that one receives or inherently carries as a result of learning the traditions of Native peoples with regard to science is important to cultivate. This includes respect for each individual's way of learning and choices for applying learning to community work and to one's life. It is essential to honor what students bring with them to school because the role of their community is also acknowledged, as well as values the Native community wishes to perpetuate.

The history of Western science relative to Native people needs to be taught. Students must be exposed to the historical context of science, and about the exploitation of Native peoples through the application of science and technology. Therefore, there is a need for a reflective and anti-racist science education.

Setting forward a process integrating Native and Western perspectives of science requires a relentless and systematic commitment to change. Efforts related to cultural research, revitalization and restoration that result in transformative change of long entrenched and culturally unresponsive patterns of science education will require patience and informed and reflective teaching practices. In addition, teaching basic academics through basic science must be an important component. These changes will also require changes in teacher education to more completely address the needs of Native students and community.

New forms of curriculum content need to be infused into the "standard" science content, metaphoric thematic areas such as "Land, Sea and Sky," "Earth, Air, Fire and Water," "Seasons of Indigenous Life," and "Sense of Place (MacIvor 1996)."

Native (i.e., Indigenous) science still exists and informs Native life in vital ways. Ways must be found to revitalize and further develop its processes through contemporary expressions of education. Native people trained in the sciences have a special role in the preservation and revitalization of Native science.

[In exploring the dimensions of Indigenous Science]…"can be found discussions of metaphysics and philosophy; the nature of space and time; the connections between language, thought and perception; mathematics and its relationship to time; the ultimate nature of reality; causality and interconnection; astronomy and the movement of time; healing; the inner nature of animals, rocks and plants; power of animation; the importance of maintaining a balanced exchange of energy; of agriculture; of genetics; of considerations of ecology; of the connection of the human being to the cosmos; and of the nature of processes of knowing (Peat, 1994: X1)".[6]

An understanding of the conflict between Western Science and traditional Native knowledge must be facilitated through research and teaching of science to Native students. Envisioning science as a Western activity with little relevancy to Native life is common among Native students. School science, Western scientific philosophy and school work all have to be understood in terms of why the conflict exists and how bridges between these knowledge systems may be constructed. Ways that such bridge building can begin to be applied include the use of exemplars from Native traditions to illustrate conventional scientific principles and highlighting the complementary nature of some traditional teachings and conventional science.

Native science evolved in relationship to places and is therefore instilled with a "sense for place." Therefore, the first frame of reference for a Native science curriculum must be the "place of the community, its environment, its history and people." Native students must be made to feel that the classroom is reflective of "their" place. Indeed, relationship to place occurs in a greater context as MacIvor states, "Respect for Native spirituality and for Native nature-wisdom embedded within it, is inseparable from respect for the dignity, human rights, and legitimate land claims of all Native peoples." Given this orientation, stewardship of place is an important part of indigenous science education. (MacIvor 1996).[7]

A transformational process must be a part of the development of contemporary science education for these students. The transformation of Indigenous education will be a direct result of the full integration of indigenous knowledge, orientation and sensibility into the teaching of science.

Many of the new genera of "creative science" methodologies which incorporate brain-based research parallel what traditional Indigenous education has always exemplified, primarily because Native American educational methods are predicated on long observation and experience with natural learning processes.

Creative Science Approaches

A number of Creative Science Strategies have emerged over the past three decades. These concepts have been spurred on by evolving research on brain hemisphericity and physiological cognitive brain functioning. Such recent creative science models— such as <u>Integrated Thematic Instruction, The Private Eye, Discovery Science</u>, and <u>(STS) Science, Technology and Society</u>—draw heavily upon research related to multiple intelligences; the inherent and complementary cognitive processes of science and art; and innate, intuitive and creative potential of human curiosity. For example, the goal of (STS) "is to empower learners to make judgments about the role of science in the daily life of their society…" (STS) assists in developing problem solving skills by analyzing situations that affect choices and human values (Bybee 1987).[8]

The Private Eye is a program which uses observations of "micro-nature" to develop a higher order of the thinking skills, creativity and scientific literacy across subjects (Ruef 1992: 15).[9] Discovery learning is hands-on, experiential learning, which helps [students] move through a continuing series of experiences that include hands-on work with science materials; it challenges them to make sense out of their discoveries through writing, library research, mastery of science vocabulary and a host of other activities that lead them to make still more discoveries (Abruscato 1996. 4th ed).[10]

The Integrated Thematic Instruction (ITI) model developed by Susan Kovalik applies brain research from over thirty years in the development of brain compatible instruction. Brain functioning is characterized by the fact that the brain has three distinctively functioning parts. Intelligence results from experiences which cause physiological growth in the brain. There are at least 8 identifiable human intelligences. The learning brain uses pattern seeking processes. Information that is not connected to real life applications is generally forgotten within a short period of time. Given these brain pattern orientations, the ITI applies the following 8 brain compatible principles of instruction to science learning:

1) Absence of threat: Learning occurs more readily in a non-threatening environment.

2) Meaningful content: Learning is enhanced by content which has personal meaning.

3) Choices: Given individual learning preferences, choices for approaching learning content enhance individual motivation to learn.

4) Adequate time: Learning is developmental and unfolds through time.

5) Enriched environment: Learning is more likely to occur in an environment that is rich in many forms of stimulation.

6) Collaboration: Human learning can be facilitated through group work and collaboration.

7) Immediate feedback: Learning is self-reinforcing through immediate feedback which adjusts learning response until success is achieved.

8) Mastery: The true evaluation of learning is the ability to use concepts and skills in real life.[11]

The work of Yupia'q educator Oscar Kwagley[12] and other like-minded Indigenous curriculum developers can be considered a reaction to the Western educational system which has consistently attempted to instill a mechanistic/linear worldview into contexts previously guided by an Indigenous worldview of relationship and reciprocity toward the natural world. This attempt to assimilate Indigenous peoples' views into the Western materialistic view of manipulation of resources is based on the premises of progress and domination of the natural world... supposedly for an improved quality of life and.....on manifest destiny as the prevailing paradigm of educating Indigenous populations. These attempts have been met for the most part by outward compliance and inward resistance by Indigenous people.

The reality is that Indigenous peoples' worldviews are about integration of spiritual, natural and human domains of existence and human interaction. Characteristics of this reality include:

1) A culturally constructed and responsive technology mediated by nature;
2) A culturally based education process constructed around myth, history, observation of nature, animals, plants and their ways of survival;
3) Use of natural materials to make tools and art, and the development of appropriate technology for surviving in one's "place;" and
4) The use of thoughtful stories and illustrative examples as a foundation for learning to "live" in a particular environment.

Various overt and covert disruptions of these traditional educational systems have led to personal, psycho-social and spiritual dysfunction we now see in Indigenous societies. Benefits of "modernity" are offset by inefficient housing, disruptions in parent - child relationships, domestic violence, suicides, alcohol/drug abuse and other forms of dysfunctional behavior. With these dysfunctions has come a general sense of powerlessness and loss of control experienced by many Indigenous people.

In short, Indigenous people are forced to live in a psychically constructed world that is not of their making or choosing and essentially does not honor who they are either personally or culturally. As Kwagley states: "Little is left in their lives to remind them of their indigenous culture; nor is there recognition of their indigenous consciousness and its application of intelligence, ingenuity, creativity, and inventiveness in making their world." (Kwagley 1995:10).

The work of Oscar Kwagley provides both a window and mirror into how culturally based indigenous science education curricula may be developed and applied. Working with traditional Yupia'q perspectives predicated on nuances in thinking, doing and learning, Kwagley has embarked upon a comprehensive curriculum development process for a contemporary Yupia'q science education.

In Kwagley's words, "From the Yupia'q person's perspective, the constellation of these new values, beliefs and practices introduced through schooling, religion, government, economics and technology represents an enormous challenge to the Yupia'q world view....so the natural place to begin is with re-establishing and reinforcing the traditional Yupia'q way of knowing and educating".

Kwagley has attempted to:
1) Examine the sum of the historical consequences of the interaction and conflict between the Yupia'q and Western worldviews;
2) Understand how people in contemporary Yupia'q communities adapt their cultural values, principles, etc. to accommodate the intersection of Yupia'q and Western worldviews in constructive ways;
3) Document contemporary Yupia'q practices in the traditional activity and setting of the traditional fish camp..(and) explore implications for the development of social, political, economic and educational institutions to the needs and aspirations of Yupia'q; and
4) Construct an epistemological framework and pedological orientation in Yupia'q traditions of knowledge as these pertain to the learning and use of scientific knowledge in a Yupia'q environment.

Oscar Kwagley's work identifies the values and life principles currently operative in the Yupia'q community and explores the extent to which existing configurations will allow the Yupia'q to reconstruct a world that will empower them with sufficient control over their own lives and give solidarity to their efforts.

PART TWO...
Footnotes

Chapter 4
1. Land, G.L. and V. A. Land (1982) present a transformational model of the creative process which is particularly suited to the development of culturally based curricula.

Chapter 5
1. Contrasting the paradigm of Western and Native American paradigms cultural expressions of science is a key aim of the curriculum model which I propose.

2. The material and research quoted pertaining to the paradigm and function of scientific textbooks in science education was paraphrased from an unpublished paper entitled, "The Function of Science Textbooks in the Thought of Thomas Kuhn," by Victor L. Austin, Santa Fe, New Mexico, 1982.

3. In The Third 1959 University of Utah Research Conference on the Identification of Scientific Talent, ed. C.W. Taylor (Salt Lake City: University of Utah Press, 1959), pp. 162-74. Reprinted in Kuhn, The Essential Tension, pp. 225-239. The phrase is used on p. 229.

Chapter 6
1. MacIvor, M.(1995). Redefining Science for Aboriginal Students. In Battiste,M.& J. Barman (Eds.). *First Nations Education in Canada: The Circle Unfolds.* Vancouver: University of British Columbia Press.

2. Hampton, Eber. "Toward a Redefinition of American Indian/ Alaska Native Education." *Canadian Journal of Native Education 20,2 (1993):216-309.*

3. Briere, E.J. (Ed.), (1979). Language Development in a Bilingual Setting. Los Angeles: National Evaluation, Dissemination and Assessment Center, California State University, Los Angeles.

4. United States Office of Education, (1997). Programs Under the Bilingual Education Act: Manual for Project Applicants and Grantees, Washington, D.C.: Author.

5. More, A.J. (1990). *Learning Styles of Native Americans and Asians.* (Report N0. RC-018-091). Vancouver, CA: University of British Columbia. (ERIC Document Reproduction Service No. ED 330 535)

6. Peat, F. David (1994). <u>Lighting the Seventh Fire</u>. Carol Publishing Group, New York (xi).

7. MacIvor (1995).

8. Bybee, Roger (1987). "Teaching About Science—Technology—Society (STS): Views of Science Educators in the United States." <u>School Science and Mathematics</u> 87 (Aprl, 1987): pp274-85.

9. Ruef, Kerry (1992). <u>The Private Eye: Looking and Thinking By Analogy</u>. The Private Eye Project, Seattle, WA, p15.

10. Abruscato, Joseph (1996). 4th Ed. <u>Teaching Children Science: A Discovery Approach</u>, Allyn & Bacon Pub, Needham Hts, MA, p38.

11. Kovalik, Susan (1994). ITI: The Model. Kovalik & Associates, Kent, WA, p255).

12. Kwagley, Oscar. (1995). *A Yupia'q Worldview: A Pathway to Ecology and Spirit.*" Prospect Heights: Waveland Press. A four year pilot project based on the thesis of this book was initiated in 1995 and funded by the National Science Foundation under the auspices of the University of Alaska, as part of the Alaska Rural Systemic Initiative in Science. It relies on the immediate environment as the classroom and incorporates Native cultural content.

PART THREE...
Looking At the Learning of Science

Chapter Seven...... Traditional Native American Education

Holistic learning and education has been an integral part of traditional Native American education and socialization until relatively recent times. Teaching and learning was a natural outcome of living in close communion with the natural world. It is only within the last three or four generations that Native Americans are experiencing "holism" as if it were new.

A major purpose of this science curriculum is to reintroduce the idea of holism and integrated learning in an interactive social environment such as the school or community. The intent is to make education for Native American students related to other learning situations and interdisciplinary activities and a more culturally interactive process than it has been in the school environment.

Traditionally, Native American teaching and learning occurred within very high-contexted social situations. The lesson and the learning of the lesson was intimately interwoven within the situation and the environment of the learner. Native Americans taught what needed to be taught in the context they thought it should be taught and at the most opportune and appropriate time. The situation provided the incentive to teach or learn, and all learning and teaching were intimately involved with daily life processes. This involved a huge amount of daily programming.

This style of learning is in direct contrast to teaching and learning as it occurs in schools in Western society. Formal teaching and learning in Western society are predominately low-context and are becoming increasingly so. Time for teaching and learning are scheduled and specific. Education occurs in the very low-context environment of the school surroundings, and interaction between peers and a few adults occurs in a highly formalized relationship. Low-context situations by their nature are very specific, and transfer of information occurs in fixed arrays and structured patterns. Low-contexts have a minimized flexibility of output of information due to the narrow perspective inherent in their structure.

High contexts are by nature open situations in which a number of different things are happening at once, at many different levels. A variety of patterns of information are transferred simultaneously. (Hall 1983: 28-29).

An example of such high context learning in contemporary society would be flight training. When learning to fly an airplane, the student is totally immersed in what is at first a bewildering array of stimuli, all accentuated by the anxieties inherent in a risk intense situation. Vocational education courses are also sometimes high context learning, sans the element of danger.

It was within high context experiences that most education occurred in earlier times. The learner's living place, as well as the learner's parents, grandparents, brothers and sisters, the extended family, the social group or clan, the community, and immediate environment provided the context as well as the tutors for learning. Every situation <u>was</u> a learning opportunity. Basic education was not separated into specific categories or in any way disassociated from such things as the natural, social or spiritual aspects of everyday living.

The goal of all such basic education was self-knowledge, "seeking life" through understanding the creative process of living, sensitivity to and awareness of the natural world, knowledge of one's role and responsibility in the social order and receptivity to the spiritual essence of the world. Because such goals required continuity of knowledge, perception, experience and wisdom, cultivation of all one's senses and creative exploration were highly valued. Verbal ability in the form of storytelling, oratory and song was highly regarded, respected only when one had something meaningful to say based upon his or her experience and place within the community.

"Formal" learning and sacred knowledge in Native American societies usually took the form of an initiation and usually occurred at graduated stages of growth and maturation. Important initiation ceremonies and accompanying concentrated education coincided with such times as the end of early childhood (6-9 years), puberty, young adulthood, middle adulthood, late adulthood, and old age. The introduction to sacred knowledge then, was informally graduated and programmed in such a way that each individual was presented with a new level of knowledge when it was felt he or she was ready to deal with it. The concept of "life-long" learning, the call word of modern adult education, was practiced by most Native American tribes long before adult education in the contemporary sense existed (Beck and Walters 1979: 84).

The contexts and mechanisms through which teaching and learning occurred included: experiential learning (learning by doing and seeing), storytelling (learning by listening and imagination), ritual/ceremony (learning through initiation), dreaming (learning through the unconscious and imagery), the tutor (learning through apprenticeship) and artistic creation (learning through creative synthesis).

Experiential learning is the most basic and the most holistic type of human learning, and is a part, in one form or another, of every Native American context and mechanism of learning. This learning requires the simultaneous "internalization" of concepts, methods and classifications that are predominately non-verbal and unconscious. The old maxim that "experience teaches" was extensively exploited in traditional Native American education. Such things as learning a traditional art form, learning to build a shelter, learning to farm or hunt, or survive in a given environment were predominately experiential in nature. Children experienced basic education through everyday work and play.

In a traditional Native American setting it was not uncommon to find children involved in all sorts of work/play activities such as building play shelters, hunting small animals, trying to catch fish, caring for babies, animals, or a small garden, making clothes, bows, baskets, pots, or making up their own songs and dances. Children were involved in the everyday work and living of their families. The fact that the skills and understandings children needed everyday were incorporated into play helped internalize this education at the child's most impressionable stage of life.

Early education helped develop and train the psycho-motor skills as well as the use of all the senses to a very high degree. A child's capacity for creative imagination and intuitive reasoning was trained by experiential learning through work/play activities. Education was completely holistic rather than just cognitively oriented.

Storytelling was both an enjoyable and very effective means of teaching and learning in Native American traditional life. Through the mechanism of storytelling, every individual was introduced to various levels of meanings, practices, concepts, ethics and codes of conduct which were meant to partially answer the "why" of the "way of the people." That is, such storytelling related the ever-evolving group life processes and introductory understanding of its members as part of a unique tribal people.[1]

Without a doubt, such storytelling was and still is a very vital way of learning. It helps to exercise such skills as memory, imagination, verbal and non-verbal communication in both the listener and storyteller.

Accomplished storytellers exploited the rich symbolism inherent in mythology and used it to its fullest potential to illuminate many aspects of human psychology and the resulting behavior therein. The hidden symbolism and use of metaphoric communication in myths of ancient peoples is only now being explored. The sophisticated nature of the storyteller's art, which employs creative use of language, evocation of imagery and theatrical ability is becoming appreciated and accepted in modern education. Native American storytellers utilized "coding" in their art of storytelling. "Coding" — the use of metaphor and symbolism within the contexts of particular stories — allowed and encouraged listeners to fully exercise their creative thoughts toward creative synthesis. They had to "read between the lines" to discover the underlying shades of meaning, concepts, or ideas. In short, they had to listen with the whole mind. "Coding" also allowed for flexibility in the *levels* of understanding that could be conveyed. The storyteller adjusted the level of meaning and symbolism to the listening audience. For younger children a very simple level of coding was utilized, for older children a slightly more sophisticated level, for teens another, and for adults yet another level was incorporated. A single story could be used to teach something to young and old alike depending on the level and way it was presented.

The creative use of wit in the stories, or in the way the storyteller presented, added to the enjoyment and "play" of the audience. Humor increased the psychological receptivity of the audience to what was being taught or conveyed. In many cases, wit in a story forced listeners to suspend their rational judgements and sense of reality and to consider the absurdity that might be a part of a story or its presentation, thereby developing a new perspective or understanding through creative synthesis.

The ability to tell and listen to stories develops a whole range of verbal and non-verbal skills, as well as what we today call right and left-brain functions, in both the storyteller and listener. This is an especially significant characteristic for teaching and learning by children. Telling or listening to stories is an almost universal activity of younger children, but it is a capacity that is rarely capitalized upon, guided, or developed toward positive learning. It is one of those subtle human activities that needs only to be exercised and valued.

<u>The tutor</u> and the master-apprentice relationship was a widely utilized form of teaching and learning in Native American society. In general this type of education took two forms.

Relationships between father and son, or grandfather and grandson, or aunt and niece, or any combination therein, constituted informal tutor-student relationships within the extended family or clan. Indeed it was within the context of these relationships that much formal and informal learning and teaching took place. Many important aspects of practical knowledge such as care for oneself, the daily work of a household, finding food, making shelter, protection and social education took place in this situation. The techniques used to teach and learn varied widely from formal lectures and demonstrations, to stories and "show and tell."

The formalized type of tutor relationship often occurred during formal training of children, young people or adults by a tribally-recognized specialist. In many Native American societies, such training was incorporated into the initiation complex of a particular tribal society or clan group. Sacred or highly specialized knowledge and training was conveyed to a chosen student or group of students.

Specialized training, such as that involved in becoming a medicine man or shaman, a priest or participant in an important tribal society, usually involved the tutor-student relationship. The relationship between the parties might last many years and occurred in either specific and highly formalized situations or very informal generalized situations.

<u>Dreaming</u> was a valued method of learning as well. Patricia Garfield, in her book <u>Creative Dreaming</u>, wrote:

"Any child who receives recognition and praise for dreaming will certainly learn to recall and use more dreams as he is rewarded for doing so

... We in the West are told that dreams are nonsense, or amusing, or psychologically revealing; accordingly we never hear the suggestion that dreams can be actively used; we do not deliberately engage ourselves in our dreams to help ourselves" (Garfield 1974: 92).

Almost without exception American Indian societies valued dreams, dreaming and imagination as very unique and powerful ways of learning, understanding, teaching and creating. The ways in which Native American cultures used dreams varied tremendously. Dreams were often an important part of religious systems because they were viewed as one of the primary means for contacting spirits to gain power and knowledge from them. Likewise, dreams were often viewed as important entities within the value structure of the social system, and because of this, the dream interpreter and dream interpretation were assigned special roles and status.

Dreamers used their dreams to deal with a variety of personal problems. Extensive use helped the individual come to terms with his/her inner self and thereby develop a more complete sense of self, more self-reliance and a greater ability to cope with stress through a process today termed "psychological immunization" or emotionally desensitizing oneself to a particular psychological situation. Dreams helped manage and treat psychological problems of individuals within a healing context by using some very sophisticated methods of psychotherapy. Native American healers helped dreamers reveal frustration, sources of anxiety and learn when to apply certain cultural curing rituals. The encouragement and value given to dreaming provided a powerful incentive to develop dream life and learn about themselves in a most intimate way.[2]

Dreamers and dreams were used to predict the future and fortune of a tribe. In this vein, many tribes and individual dreamers developed rituals which were meant to cultivate this capacity. Traditionally their insights were considered keys to success in life. In a variety of ways Native Americans taught the value of a helpful dream to all tribal members from the youngest to the oldest. The more importance one gave to dreaming at all ages the more relevance dreams would have to the waking life of all the people. Many Native Americans believed one could eventually plan the content of dreams for a specific purpose. Indeed, many Native Americans were reputed to have developed the ability to control their dreams and to produce very specific elements.

Cultural rewards and high status rewarded those whose dreams helped people in some way. Native American cultures provided psychological rewards for "the cultural-pattern dream." By doing this, the special place of dreaming in the fabric of their societies was powerfully reinforced.

In addition to "dreaming to help the people," Native Americans sought to catch hold of their deepest selves and create meaningful personal or group rituals, ceremonies or customs. They immersed themselves in their dreams so they were able to give their dream symbols some creative waking form. Much of Native American art, song, dance, story, mythology, poetry, ritual, and ceremony is a direct result of this application of dreaming.

Ritual and ceremony infused every aspect of traditional Native American life. The spirit and the spiritual were at the center of each human being and all that made up the universe. Through ritual and ceremony, teaching and cultivation of the spirit and the spiritual was engendered, from very simple symbolic acts recognizing one's relationship to the spiritual to highly organized, high-contextual events which involved all the people of a particular tribe for several days.

A ritual's main purposes were to provide a focus of reflection upon the great mystery in one or more of its manifestations and to help revitalize the individual and the group connection to themselves and the world.

Through the symbolic use of prayer, song, dance and communal activity, Native Americans developed highly creative techniques for guiding social behavior and ethics. The social psychology inherent in ritual and ceremony provided powerful group empathy and cohesion which reinforced the social self-image of each individual participant.

Ritual and ceremony formed a major foundation for the socialization of children in all Native American cultures. The important initiations at different stages of the individual's life cycle helped him/her internalize the knowledge inherent in each initiation activity. Every individual needed to be in touch with the spiritual throughout his/her life if he or she was to live a whole life. In every Native American language there is a phrase which is said or implied in ritual and ceremony. This phrase has variations from tribe to tribe, however it is usually translated to mean "to find our life" or "in search of our life."

For Native Americans, this phrase describes the main goal or quest of learning and teaching. **Indeed, "to find life" and understand its manifestations within each individual self is the ultimate goal of all education.**

In ritual and ceremony, participants recreated an important part of their historic selves and in a metaphoric sense retraced their steps as a social group of people in process. In Jungian terms, this recreation and reenactment of the cultural creation myth allowed for the expression of the mythical self and all the archetypes therein. The mythical self was fully realized as an integrator of "the spirit within" by tribal societies and played an important role in traditional Native American education.

These general areas of traditional Native American methods of education underlie the curriculum model presented in this work. Because Native American cultures have changed a great deal since the time when these methods were extensively used and because modern schooling has usurped their functions and the contexts in which they were applied, many of these methods are no longer as viable as they once were. This does not mean, however, that they are not highly effective teaching methods. They can be adapted in a variety of ways to modern educational contexts and teaching requirements, as they parallel many of the new methods espoused in holistic education. Their synthesis and application present both a challenge and a vital component of relevancy for the educator of Native Americans. What is required is an openness to the methods and an understanding of their role in the total educational process.

PART THREE...

Chapter Eight......Science Learning and the Creative Process

Because of its unique nature, creativity does not fit comfortably into any structured theory which attempts to define it or reduce it to a list of traits. For many creative artists and scientists, creativity is simply the way that they do their work. It is that mysterious yet illuminating process of thought which helps them see part of their reality differently. It is a type of looking and seeing which allows them to restructure a part of their reality in new and innovative ways.

Some scholars of creativity, like Edward De Bono, reject the notion that there is any sort of transcendental mystique to creativity. De Bono believes that creativity is essentially a high level thinking skill which can be taught by helping people to "take unorthodox leaps of imagination." Intelligence is a potential while creative thinking skills involve working out an overall design toward a general goal. "Lateral thinking," or creative thinking, involves the "ability to change perceptions and concepts by cutting across patterns...."Creativity is a function of motivation — curiosity, wanting to do things differently — and talent...largely a method of thinking, style, pattern, habits, and techniques which can be internalized." (OMNI 1985: 75-76, 116-120).

In this chapter, creativity will be explored in reference to three perspectives:
    1) its relationship to cultural learning and adaptation,
    2) its interplay with human intelligence and symbolization, and
    3) science learning.

Creativity has always been an integral part of cultural adaptation; each generation must redefine itself within cultural contexts. In this complex process each adapting generation must resynthesize the old with the new. Culture, after all, involves constant learning, and at times, the creation of new "grammars," or new ways to talk about the old and the new. At other times, it requires the restructuring of old frames of reference, or developing entirely new frames of reference. New meanings evolve for old symbols and old relationships. Through this process of redefinition, the cultural mindset evolves.

The creative dimensions of the adaptation and assimilation lie in the continual evolution of mindsets, ways of looking and seeing which allow for transformation of symbol systems within a culture to higher levels of meaning.[1]

Each culture encounters and represents reality through the creative forms of science and art. The key to understanding the cultural creativity of any society may very well lie in understanding of these expressions of "reality." One possible way to initiate creative learning is to view reality as a cultural process of meaningful symbolic activity. C.A. van Peursen writes,
"Human history could be compared with a continuously changing chessboard. Man develops his strategies and gains in this game his own identity. The adversary is only visible through the symbolic interpretations. Religion, sciences, the arts, social developments, daily experiences, all this is to be found in this field of symbols" (C.A. van Peursen 1981: 177).

Creativity at every level of expression involves the development of new strategies for problem solving, which in turn entails the development of new grammars. Based on a culture's unique configuration—its history and its perception of what it opposes and what it wants to become—creativity becomes a survival strategy.

Creative strategizing is also an important part of the development of educational curricula. A strategy which underlies the presentation of science to Native Americans must, then, involve the development of a biculturally relevant grammar which will include the symbolic systems utilized within both Native American and Western cultural sciences.

If culture is an active learning process, the interaction between symbolization and meaning-making on the one hand and the creative response to a particular environment on the other becomes important. Science is a cultural and symbolic representation of reality because of our evolved set of beliefs. In turn, creative responses transform the culture. The process is cyclic within the parameters set by culture and environment, and Native American cultures are no exception to this process. Any explanation of a Native American cultural group's way of knowing and relating to nature must necessarily begin with an exploration of that group's creative strategizing.

In order to gain a more complete understanding of creativity and cultural learning, it is important to explore the interrelationship between creativity, intelligence and symbolization addressed in the work of Howard Gardner. Gardner defines intelligence as "the ability to solve problems or to create products which are valued in one or more cultural settings." Based on his research on the development of the artistic and cognitive abilities of normal and gifted children, Gardner suggests that there are at least seven domains of intelligence, many more than had been formerly acknowledged by other researchers. **These are linguistic, logical/mathematical, spatial, kinesthetic, musical, interpersonal (understanding of people), and intrapersonal (understanding of self)** (Scherer 1985: 31).

In his Theory of Multiple Intelligences, Gardner contends that the brain perceives information based on predetermined tendencies which have been conditioned through the interplay of individual, environmental and cultural factors. Unlike Piaget's developmental/operations theory which implies that mental faculties such as learning, memory and perception can be applied equally well to any content, Gardner suggests that each domain of intelligence relies upon its own characteristic mental operations.

Cultures emphasize development of competencies that are deemed valuable within that context. Among many Native American groups, interpersonal intelligence along with certain expressions of kinesthetic, musical and linguistic intelligences are valued. Leadership, the ability to sing and dance well, the ability to use one's native language eloquently, to get along with people well, and athletic prowess, in earlier times associated with the development of warriors, are valued competencies. Traditional expressions of these intelligences have changed but they are still valued as important individual/cultural characteristics.

An intimate interrelationship exists between artistic and scientific thought processes, asserts Herbert Read in his classic book Education through Art. Read illustrates the modes, both conscious and unconscious, of the integrating of creative thought in art into science, and then of science into art. He emphasizes that, **"art is the representation, science the explanation of the same reality."** Implicit in this idea is the notion that art and science are modes of perception which are not opposed to each other but complementary and dependent upon one another. This relationship becomes obvious when one explores Native American art and its symbolic reflection of ideas about nature (Read 1945).

Howard Gardner extends Read's thesis considerably by incorporating the most recent insights gained from cognitive psychology, brain research and creativity. Gardner's studies illuminated, from a cognitive psychological standpoint, the characteristic mental processes involved in thinking, problem solving and creating. His belief that "the key to various forms of creation lies in the understanding of how humans use symbol systems" underlines the importance of research involving traditional Native American forms of symbolization of nature (Gardner 1981: 300-302).

According to Gardner, the logical/mathematical domain of intelligence is one which develops rapidly and declines rapidly. Other intelligences develop sporadically or evolve slowly over an individual's life span and may or may not manifest depending on whether or not the interplay of circumstances allows for their expression. Gardner's implication is that there may well be a period when a subject such as science can be taught most effectively based on when a domain, such as logical/mathematical intelligence, is at its highest potential. Spurts of creativity in some individuals do seem to coincide with the periods when certain types of intelligences are "pregnant" with potential. Creativity within this frame of reference may be said to syncretize with a specific domain of intelligence.

Gardner believes that modern schools diminish the importance of all other domains of intelligence in favor of logical/mathematical intelligence. Yet even in science and math there is an interplay of the other intelligences, such as kinesthetic, spatial, interpersonal and linguistic, that are essential in the development of a holistic perspective (Scherer 1985: 32).

Gardner, like many researchers before him, emphasizes symbols as mediators of cognitive processes. Areas in which symbolism plays a key role, such as language, art, science, and religion, are now being reexamined in the light of the findings in brain research and creativity.

What is a symbol? Very simply, a symbol can be said to be any entity (material or abstract) that can denote or refer to any other entity. However, the dynamic nature of what symbols do is anything but simple. For instance, symbols can represent various levels of human feeling, or they can come together to form complex symbolic systems such as a language, an art form, or a cultural science. They can become, when combined, self-contained creative products such as myths, rituals, poems, scientific constructs, artistic creations, or a host of other conveyances of meaning

(Gardner 1983: 301).

Education is the learning of culturally-relevant symbol systems. Every educational endeavor involves to some degree or another the internalization of symbol systems. Gardner, in his review of the process of symbolization, divides its development into the four human developmental phases beginning with infancy, then on to early childhood, late childhood and adolescence.

In infancy one begins to see basic expressions of the seven domains of human intelligences and basic utilization of symbolic recognition. During early childhood (two to five years) there is a phenomenal development of symbolization in which children develop the basic ability to use language, dance, play, sing, use tools and create art. In late childhood and adolescence, children develop a fuller mastery of symbolization at a time when culture becomes a key factor(Gardner 1983: 303-310). In later stages of development, intelligence and symbol usage are channeled and conditioned by cultural requirements. Symbolism is extremely significant in an understanding of the overall mindset of a Native American encounter with natural reality.

Creativity in science learning is as diverse as phenomena to be found in nature. As there are patterns to be found in nature, there are likewise patterns discernible within the creative thought processes inherent in the learning of science.

The observation, thinking processes and problem-solving strategies used in the discipline of "nature watching," the basis of all scientific investigation, require the creative process in oneself and in others. Watching nature, one sees that creativity is continually illustrated in aesthetic design and cyclic patterns in natural phenomena. All plants and animals adapt to their environment creatively. Creativity in science, therefore, has an "inside" and an "outside" dimension. Insights into the nature of creativity can be derived from the internal perspective of the learning observer (the inside) and in the natural phenomena being observed (the outside). In the learning and teaching of science the inside perspective necessarily becomes the key focus of attention. Nature is, after all, her own best spokeswoman.

Creativity seems to follow a discernible pattern, which coincides with the development of learning. That is, it moves from learning the basics to practicing what has been learned, to normalizing or internalizing what has been learned, to taking apart what has been learned and restructuring it, to synthesizing and transforming what has been learned to a new and higher level of understanding where the cycle begins again. Creative seeds of thought and activity which flower and bear fruit always seem to cross-fertilize to other situations, thoughts, or disciplines.

Creativity seems engendered through a process of searching, seeing, then incubating in reference to a task. This generally begins with trying to make sense of a paradox, which leads to observation of a pattern in an attempt to bring together the dissimilar. Incubating involves the conscious and unconscious "trying out of possibilities," after a puzzling thought or observation, which leads to more observation followed by manipulation and "play" in the search for a possible answer.

In any learning situation creativity must involve immersion in an activity of relevance for an individual. Whether or not an activity is relevant is highly dependent on personal and cultural factors. The social and cultural environment is always a factor in motivating and engendering creativity as there is an enormous investment of self in the process of creating.

Finally, creativity at its higher levels requires the learning of a body of content, learning to see from a scientific perspective. "Seeing" involves the development of visual thinking in tandem with verbal/logical modes of thinking. Creativity can be likened to a seed full of potential which to reach its potential must grow through an outward process of sprouting, developing leaves, flowering and bearing fruit. This outward growth, visualization and verbalization, makes explicit what is implicit in a creative thought.

Scientists rely heavily on a creative internal dialogue as they work through a problem. Through the use of journals and "notes of the mind," scientists have conveyed their internal dialogue and much of what we know about the creative thought process. The naturalist journals of Thoreau, Emerson and Parkman are examples of creative verbal thinking. Creative writing becomes an integral part of the creative process in science since it allows for seeing and feeling when encountering and learning about natural phenomena.

Not all thoughts lend themselves easily to verbalization. Seeds of thought generally begin with an image, a vision of a pattern, a visual intuition which is the result of "seeing" something either internally in the mind's eye or externally in the environment. The work of Susanne Langer, Ernst Cassier, Edward Sapir and Benjamin Whorf characterize the role of language and verbal thinking in the naming and categorizing of human thought in reference to the inner and outer environment. For these scholars, language is thought and all forms of creativity are highly dependent on the process of verbalization in some form. (Steiner 1985).

Visual thinking usually precedes verbal thinking and verbalization in the creative process. In a sense, verbal thinking serves to explain visual thinking. Throughout the creative process these two modes of thinking are intimately intertwined.

Visualization in science may consist of a kind of weaving of intuition, observations and concepts in the mind's eye. To use an artistic analogy, this process may begin with a visual rough sketch of the area of concern, followed by a play with composition. Applying tools of thought, which have been learned in order to fill in details, further defines forms, colors and texture. Visualization allows you to place the problem in context before you so you can look at it, study its nature and in doing so learn to ask the right questions.

Creative scientists such as Einstein and Heisenburg have suggested that the ability to ask the "right question" is more preeminent in approaching a problem than prior knowledge. Intimately intertwined with this ability to ask the right questions is the ability to visualize the problem in a field of possible questions. One becomes a psychological participant in the holographic visualization.

Visual thinking has a logic which can best be exemplified through the process of drawing. There is a direct connection between drawing and seeing. Just as the application of the scientific method can exemplify certain aspects of the creative process, likewise, the act of drawing can be an exercise in creative visualization.

Drawing is a thinking tool. It sharpens one's ability to perceive and observe and may well be one of the simplest ways to develop visual thinking. Drawing a problem or concept in science can be of tremendous value towards the goal of learning to see scientifically.[2]

Much like listening, drawing focuses concentration on a task and forces close observation. The process of drawing always involves looking at what is being drawn in many different ways which elicit the use of analogy and metaphor. This involves an affective relationship tied to the motivation to pursue an interest further.

A basic perceptual skill for conveying and understanding meanings and thoughts, drawing is similar to reading and writing. The discipline involved in drawing is the very same discipline of observation that is required in learning a system such as science. Science involves learning how to observe, and then how to ask the "right question."

The visually drawn analogy, like the verbal analogy, condenses an entire thought toward further insight requires that students develop the ability to make knowledgeable connections between what they know and see that is similar in a problem with something from their own experience. At the collective level, Native American mythology is replete with just such analogies, which are both symbolically drawn and verbalized. These analogies contain dozens of insights concerning natural processes and natural phenomena which analytic description has difficulty in relating.

All children exhibit some affection toward and curiosity about their environment. Within certain cultures, such as that of Native Americans, respect for nature- its exploration and symbolization- is intimately interwoven in the fabric of cultural expression. An affection (biophilic sensibility) for nature is transmitted to children through informal learning. Affective thinking—feeling—then influences learning about nature.

All learning contains an affective component related to motivation; traditional Native American teaching and learning relies on it. The Hopi, for example, go to great lengths to engender a positive attitude toward what must be learned by young children. Dorothy Eggan described traditional Hopi schooling and its role in maintaining Hopi society. Cultural continuity among the Hopi, Eggan hypothesized, might well be dependent upon strong emotional or affective conditioning during learning which is integrated into social acts within the culture (Spindler/Eggan 1963: 324-325 & 345).

The expression of science is highly dependent on the affective domain of learning. A love for the natural world through experiencing it engenders love of learning about nature. When a child feels the natural world through seeing, touching, smelling, even tasting, the seed of creativity is planted. When his/her curiosity and his excitement about discovery and finding out how something works, is encouraged, the creative synthesis of visual and verbal thinking is facilitated.

The question is how to provide models of thinking and approaching a problem which build on the disciplines or cultures and the accumulation of knowledge yet allow for individual expressions of thought and action. Much of what is taught as science tends to lose students in detail, which leads to alienation.

Through the mentoring relationship, creativity in the student is greatly stimulated. In this affective and informal process, creative ability grows through the enhancement of visual and verbal thinking, the greater understanding of basic tools and concepts and a variety of other skills pertinent to a discipline of observation such as science.

Recognition and positive experience within such relationships has been shown to influence a student's sense of direction, motivation to learn the basic tools required in a particular discipline, and his sense of participating in a greater whole.

Characteristics predisposing the development of creativity in which affective learning plays a significant role include:
1) An early interest in explorations of nature;
2) A sense of wonder and interest in going beyond the surface of things;
3) A playfulness in reference to ideas and approaches;
4) An interest in finding and exploring patterns in nature;
5) A family which encourages the development of a child's creative interests.
6) The focused interest and encouragement of a significant other, such as a parent, teacher, relative or mentor;
7) A community of friends providing an exchange and collaboration of ideas; and
8) Confidence in working through mistakes in creative problem-solving.[3]

An important variable is the role of a community of peers, of "like minds." Human beings have always learned a great deal from each other while interacting in group endeavors. Students actually seek out those of similar interests and motivation. Also, encouraging interdisciplinary perspectives of a natural phenomena through the lenses of art, psychology, history and other subject areas helps students gain a broader perspective of what they are studying.

Finally, the role of the teacher in facilitating creativity in science learning is enormously important. Since science is a formalized system of observation and content, much science learning must necessarily involve formalized teaching. The teacher becomes, in many respects, the key to the creative synthesis of the basic principles and content of science with a student's innate intelligence, attitudes, and motivation.

How then can a teacher facilitate creativity in science? One way this is done is by dramatizing and demonstrating the visualizing/verbalizing process through example, through modeling, and through involving students in exercises which allow them to see a problem, a question, or a situation from many perspectives, and then to visualize and verbalize what they see in a creative/descriptive form.

Second, by encouraging the application of analogy in the process of science learning. This can be done by exercises in which both teacher and student use analogy in the presentation and learning of science concepts.

Third, by allowing students the opportunity to play with ideas and concepts in science through constructing their own creative expressions of those concepts. Through actual involvement in the creative process, students learn how to explore the parameters of a concept, to explore relationships in a problem, to weigh the consequences inherent in the problem, and to learn how to see wholes as well as parts of a problem.

Through these activities, teachers help students to realize that science is a creative process, that insights are gained through combining and exercising many human abilities, and that this is the nature of "science-in-the-making."

In summary, what I have attempted to touch upon are different dimensions of the creative instinct inherent in the holistic expression of human cognitive growth and development.

Science is first, last, and always a creative process and should be taught as such. Science is not merely a body of facts which must be memorized. Its learning involves the development of creative potentials and the cognitive abilities of logic, observation and evaluation.

Science always involves holistic thinking. Although science learning primarily develops logical thinking abilities, it also relies heavily upon intuitive thinking and nonlinear imagining abilities. The work of Jerome Gothenburg sheds light on two dimensions of creative thinking, intimately intertwined with the science thought process, which he calls "Janusian thinking" and "Homo spatial thinking." Janusian thinking is defined as "actively conceiving of two or more antithetical ideas, images or concepts simultaneously." In this type of thinking, opposites are thought of as being equally valid, thinking integral to creativity and intellectually pragmatic as well as artistically intuitive. Homo spatial thinking is defined as "actively conceiving two distinct entities in the same space," thinking is reflected throughout the creative process and primarily visual in nature. Homo spatial thinking is a mode of abstracting which "affects the filling of gaps and the formation of wholes." Both types of thinking are important to science as creative scientific discovery is often the result of going beyond the parameters of what is known. (Rothenburg 1982: 61-71).

Science is symbolic meaning-making in reference to the natural world. Classification systems, conceptual frameworks, theories, hypothesis formation and experimentation all involve the development and application of symbols as tools toward the goal of arriving at a meaningful understanding of the process of nature. Symbol systems guide and develop our perceptions and relationships not only of the natural phenomena to which they are applied but also to the knowledge which is accumulated through their application. Symbols and the systems of which they are a part are always influenced by culture at the collective level. Symbols provide the grammars of cognitive thought and every cultural system of science has its own grammar. The grammars must be learned and then resynthesized through creative activity if students are to become truly "literate" in the language, thought and application of science as a cognitive tool.

Science is highly kinesthetic in that it involves learning through encountering, experiencing and experimenting. The kinesthetic aspects of learning science are not unlike those involved in artistic creation. The development of hands-on, experiential learning which incorporates the kinesthetic aspects of science must be given a high priority in the development of science curricula. Science mirrors art at the creative, cognitive, and kinesthetic levels. As such, art can play a vital role as a mediator in the learning and teaching of science.

Science always involves problem-solving strategies. These strategies are always tied to the unique context, characteristics and requirements of the problem which is in turn directly related to a certain level of cognitive or learning development. This means that, ideally, the problem to be solved must be closely matched to the relative phase of cognitive development of a student. It also means that science education should place great emphasis on problem and project orientations which are closely matched to students' capacities. Tailor-made science curricula must become the norm rather than the exception, especially for Native American learners.

In conclusion, science is at once a creative process, a culturally-defined expression and a problem-solving strategy. Science curriculum development must begin to more completely address what is currently known about each of these related processes. Each aspect discussed presents far-reaching implications for the learning of science. Creativity in terms of science curriculum development is both a challenge and a necessity for the future of science education.

PART THREE...

Chapter Nine......*Ethnoscience, A Native American Perspective*

Ethnoscience, once isolated in the scholarly domain of anthropology, has taken on new dimensions with the increasing influence of linguistic and general systems theory, humanistic psychology, quantum physics and holism. Essentially ethnoscience is the study of a process by which every cultural group develops strategies to make nature accessible to reasoned inquiry. As a basic social thought process and way of adapting to a natural environment, it is unique to each culture, yet it reflects a common breeding ground between all cultures — the human mind. In this respect, all ethnosciences, including that of Western civilization, seek relatively "objective knowledge of the universe; all proceed by ordering, classifying and systematizing information; all create coherent, internally consistent systems, and all systems appear to be based on similar types of mental observation" (Cole, Gay, Glick 1978: 6).

Ethnoscience reflects the characteristics of a particular culture; its implications, however, reach much farther. In studying the ethnoscience of any group, one begins to develop intuitive insights concerning that group's way of living, perceiving, learning and acting in relationship to their natural environment. Nowhere is the path to understanding the whole as direct as through the ethnoscience of a culture.

For this discussion, ethnoscience will be described as: "the methods, thought processes, mindsets, values, concepts and experiences by which Native American groups understand, reflect and obtain empirical knowledge about the natural world."

There are differences among cultures in the interpretation, application and exploitation of knowledge gained through this study. In the cultural perspectives of science they form the basis for classification. Native American cultures generally classify natural phenomena based on characteristics which are readily apparent or experienced; such classifications are based on a high degree of intuitive thought. Western science, in contrast, "relies more on properties that are inferred from necessary relations in the structure of the objects classified" (Cole, Gay, Glick 1978: 8).

It is in these differences between cultures that the most basic problems in knowledge transfer and mutual understanding occur. However, science, as a cultural extension of the human race, is a tool for obtaining and relating understandings about natural realities. Based on this proposition, the study of the ethnoscience of Native American cultures becomes a valuable tool for understanding not only the cultural influences in science as a whole, but a way for Indians and non-Indians alike to gain valuable insights about themselves and the unconscious cultural conditioning of their perspectives of natural reality.

Hollowell, one of the early advocates of the study of ethnoscience wrote:

"What I wish to develop is a frame of reference by means of which it may be possible to view the individual in another society in the psychological perspective which his culture constitutes for him ... rather than content ourselves with the perspective of an outside observer who even prides himself on his objectivity. All cultures provide an orientation to the world in which man is compelled to act. A culturally constituted worldview ... by means of beliefs, available knowledge and language mediates personal adjustment to the world through such psychological processes as perceiving, recognizing, conceiving, judging and reasoning. It is a blueprint for meaningful objects and events ..." (1955, 1963, quoted in Black, 1967: 12-13).

The knowledge of every Native American culture is directly dependent upon belief systems, attitudes toward nature and actual experiential interactions with nature. The meanings which result from this cultural interaction with the natural environment constitute the foundations for Native American cultural sciences.

Hollowell (1960) discussing the Ojibwa Indians, gives the following important illustration of the nature of such knowledge.

"Thunder birds and monster snakes ... are important items in the behavioral environment of these Indians. Since from our view thunder is a part of our physical environment and monster snakes are not, we might be inclined to make a distinction between them. But if we do this we are making 'our' categories a point of departure." (Quotation in Black 1968: 12-13).

The ethnoscience movement has generated substantial research concerning Native American cultures in such areas as ethno-botany,. ethno-pharmacology, ethno-zoology, ethno-medicine, ethno-psychiatry, ethno-entomology, and ethno-astronomy. All such studies have attempted to present the Native perspective of cultural science.

In summary, the strength of ethnoscience as a methodology lies in its systematic attempt to understand a cultural system from within the system based on the perspective and criteria of that system. This proves valuable in the development and presentation of culturally-based curricula. In addition, this study allows for the gaining of perspectives of human learning behavior that are based on both conscious and unconscious cultural processes. The approach lends itself to the view that culture is communication and that education is essentially a system of communication. This allows for the development of a systematic exploration of the worldview which underlies every cultural science. It also focuses upon the need to view science as a form of communication complete with a "particular kind of language" and specific content.

Expressions of science thought process in traditional Native American culture have ranged from the simple practical technologies developed to survive in a given environment to highly complex and elaborate technologies developed by the "high" civilizations of Mexico, Central and South America. These expressions of the science thought process have all taken distinctive cultural forms which reflect the way each Native American group has adapted to a particular place and environment. This process has been reflected in Native American agriculture, medical practices, astronomy, art, ecological practices, hunting and gathering.

There is no word in any traditional Native American language which can be translated to mean "science" as it is viewed in modern Western society. Rather, the thought process of science — which includes rational observation of natural phenomena, classification, problem solving, the use of symbol systems and applications of technical knowledge — was integrated with all other aspects of Native American cultural organizations. Therefore, in order to begin to develop an understanding of Native American cultural sciences, it is necessary to look at the reflections of the science thought process in such areas as the epistemology inherent in selected Native American cultural philosophies, ecological practices, arts and mythology. In each area relative differences between Native American and Western mindsets are indeed great.

Many of the principles of Western science are based on a type of logic and mindset which requires hierarchical thinking. Non-reciprocal causality, for instance, requires that one think of phenomena in the following way: "that for every effect there is one single cause which can be objectively observed and described given the proper tools, the correct hypothesis and appropriate experimentation." Non-reciprocal logic, or what has been popularly termed "linear thinking," conditions for "mono-polarization" in both thinking and personality development. "Mono-polarization" is defined as "a psychological need to believe that there is one universal truth, and to seek out, find security in, and hang onto one authority, one theory, uniformity, homogeneity and standardization"(Maruyama 1978: 467-68).

This mindset of "non-reciprocal causality" is one of the unconscious dimensions inherent in the cultural mindset of Western science and is reflected in the way science is presented in American schools.

That is, as presented and internalized there is "only" one way to view nature scientifically. But there are, in reality, a number of dimensions to science as a thought process, which, in turn, have evolved into a number of different, yet internally consistent, ways of viewing nature. The ethnosciences of American cultures represent some of these alternative ways of perceiving and interacting with nature.

For example, the mutualistic logic and orientation to "reciprocal causality," (the notion that cause and effect in nature have a reciprocal effect on each other; that cause and effect cannot be isolated from other causes and effects with which they share a holistic relationship within a system), is illustrated by the concept and expression of mutualism among the Navajo people (Muruyama 1978: 460-462).[1]

The mutual relationship between all things in the natural world — animals, plants, humans, celestial bodies, spirits, and natural forces — forms the basis for traditional Navajo concepts of the universe. The Navajo believe that natural phenomena can be manipulated by humans through the application of the appropriate practical and ritualistic knowledge. In turn, natural phenomena, forces and other living things can affect humans in a number of ways. Therefore everything affects everything else through a complex web of interrelationships. Everything that is a part of the natural world has distinctive qualities which can be dangerous or beneficial to a human, depending on his or her behavior toward these natural entities.

The maintenance and perpetuation of harmonious interrelationship with elements in the natural environment is the ideal goal of human behavior. Harmony is viewed not as a static state of being, but as a dynamic and multi-dimensional balancing of interrelationship. When disharmony occurs, every attempt is made to restore harmony.

Knowledge of how to maintain this dynamic harmony of relationships is considered of utmost importance. An understanding of the complex nature of natural forces and their "ways" of interrelationship is essential to the well-being of all. In Navajo culture, as with all other Native American cultures, there is no separation of science, art, religion, philosophy and aesthetics. All of these "cognitive maps" of human culture are mutualistic and remain integrated in all aspects of traditional Navajo culture.

The Navajo ceremonial "sing" is based on a "chantway," a complex of songs, prayers and actions designed to restore interrelationships which have lost mutualistic harmony with each other, resulting in a specific illness of either an individual or the group. A "sing" represents, for the Navajo, mutualistic interrelationships with each other and with the phenomena and forces of nature. It involves the cooperative efforts of many individuals and reaffirms the cultural mindset of Navajo mutualism (Muruyama 1978: 462).

In tracing the causes of an illness to its sources, the Navajo medicine man attempts to affect the cure of the patient through the application of his knowledge of interrelationships. Sand painting, elaborate prayer and ritual and prescriptions based on understanding the illness and specialized medical knowledge combine to restore the loss of harmony. The cure also illustrates the integration of science, art, and religion in Native American cultures. All who take part in these activities lend themselves to a cooperative effort in restoring harmony both for the patient and themselves. During the feasting, dancing, singing and other social activities, the mutualistic orientation of Navajo thinking is psychologically and concretely expressed and confirmed (Muruyama 1978: 462).

Other cross-cultural examples of mutualism are represented by the Chinese expressions of "complementarism" through their focus on the complementary interrelationships between "yin" and "yang," the two primordial natures of the universe; and the Japanese expressions of "situationalism" in which combinations of individuals or natural forces define the appropriate actions. Expressions similar to these examples of mutualism can be found in all traditional Native American cultures. Mutualistic logic has also become an integral part, a pronounced departure from mainstream logic, of modern Western scientific disciplines involved with "general systems theory" and "quantum mechanics."

Mutualism with its various expressions in Native American ethnosciences, ecological practices and arts, orients presentation of bicultural science. In addition, comparison of the process and event orientation of Native American ethnoscience with the objective, conceptual orientation of Western science presents a dynamic way to show these ideas of science as two very different thought processes(Muruyama 1978: 468).

Ecological practices of Native American tribes also demonstrate the science thought process in Native American culture. The traditional attitude of Native Americans toward their natural environments reflected a pervasive and deeply internalized appreciation for nature and its aesthetic beauty. Native Americans were America's first ecologists, and being in harmony with their natural surroundings was an integral part of lifestyles. Misuse or bad behavior toward natural entities, which led to their destruction or despoilment, was considered sacrilegious and exacted strong penalties. The sacredness of nature and the various religious taboos associated with the disrespect of such natural entities as plants, animals, wind, water, or the earth itself have often been viewed as "superstition" and categorized as expressions of animism or pantheism. However these categorizations reflect only a superficial perception of what in reality was an outward expression of a highly developed ecological mindset. The sacredness with which they hold the natural environment is the unifying thread all Native American cultures possess in common. Respect and conservation of natural resources was a vital social ethic since the very survival of the group often depended upon the maintenance of these sources of life (Hughes 1984).

The complex rituals associated with the growing of corn and the coming of rain in Southwestern Pueblo groups illustrate not only this ecological ethic, but also an understanding of their relationship to these natural entities and the land itself. **Every tribal group evolved their knowledge of nature around the central theme of man as part of his environment, not its master.**

The so-called totems and spirits with which all Native Americans symbolized their relationship to their world have often been misunderstood. In reality the ecological relationships, the sacredness of nature and the constant "seeking of life" are underlying mindsets focused upon in Native American ritual. Fetishes and other paraphernalia which are present in many Native American rituals were highly respected because they were symbols that represented the sacredness of the various forces of nature.[2]

The view of nature as a spiritual reality directly affected Native American science process and associated technologies. In general, Native American concepts of nature were not meant as explicit explanations of natural processes as are concepts in Western science. Rather, concepts such as animal or plant spirits, benevolent or malevolent forces of nature, and the mythological or ritualistic symbolic representations of nature, were symbolic representations of essences and relationships which Native American groups have come to understand through generations of experiences within a given natural environment. These concepts and symbolic representations reflected a highly evolved resonance—a feeling for the natural environment which Native Americans shared so intimately that it was commonly accepted that it was possible for humans, other living and natural forces to communicate with and affect each other through their interdependencies and reciprocal relationships.[3]

"The Indian view of nature comes from deeper inside the human psyche than mere rational thought or intellectual curiosity, although Indians certainly have these too. But Indians regarded things in nature as spiritual beings, not because they were seeking some explanation for natural phenomena, but because human beings experience a spiritual resonance in nature" (Hughes 1984: 16).

Because of this resonant relationship with nature, Native American tribes developed ritualistic expressions around the recognition, celebration and evocation of mutualism with the natural environment. Whether it was a Pueblo dance for rain, the hunting of game, the planting of corn or the healing of the sick, Native American rituals sought to maintain the harmony of these relationships and through this "seeking of life" gained a glimpse of the sacred whole of which they were a part. Native American ethics concerning the natural environment were geared toward the preservation and perpetuation of all life. Everything in nature was imbued with a spirit which was a part of the "Great Mystery" and therefore, was also a part of oneself which had to be respected.[4]

The "Great Mystery" breathed life into everything; therefore, all natural phenomena had the power to affect everything else. This was especially true for such things as wind, water, fire, lightning, the sun, moon, stars, and certain birds, animals, and plants. In addition, everything in nature was viewed as having intrinsic value and therefore, could not be exploited simply for the sake of exploitation without dire consequences. Traditionally, this understanding of mutual interrelationships was not merely a philosophical concept. Native Americans lived this interrelationship in their adaptation and interaction with the natural environment. In short, Native American cultural sciences were sciences based on experience and a high level of sensitivity and intuitive insight which is only now being explored in modern Western scientific philosophy.

Traditionally, there was no specific word for "art" in Native American languages. Native American cultures viewed the creation of art as a natural way to communicate their perceptions of nature and their feelings and interrelationships with different natural entities within their environment. Familiar images within nature were incorporated into designs of Native American art. Nature provided the Native American artists with inexhaustible content for creative expression. All Native American art forms — from pottery, jewelry and weaving, to stone sculpture and architecture — provided mediums for expressing their maker's perception of natural phenomena. Clouds, birds, animals, fish, wind, water, sun, moon, insects, plants and spirits represented mutual relationships among all things. Each traditional art form required the learning and mastery of particular types of technology. For instance, certain forms of pottery such as that of the Rio Grande Pueblos of New Mexico require great skill and a substantial knowledge of the nature of various kinds of clays, slip and pigmentation characteristics, preparation and firing techniques. Weaving, basket making and architecture all required great skill and a high level of knowledge of the nature of the materials used.

Expressions of "resonance" with the natural world required the application of material technology, creativity and problem-solving skills, with the same kind of processes used to calculate, for instance, the movements of the sun and moon, the development of healing techniques and successful hunting practices, all of which required the application of a basic understanding of natural entities and the science thought-process.

PART THREE...

Chapter Ten......Indigenous Science: An Overview

What is Indigenous science and what is its connection to the emergence of a contemporary Indigenous epistemology of education? In order to address this question it is important to first define the parameters of the "indigenous science" paradigm as it is currently viewed. In fact, Indigenous science is a broad category that includes everything from metaphysics to philosophy to various practical technologies practiced by Indigenous peoples both past and present. At its most inclusive definition "Indigenous science" may be said to include practically all of human invention before the advent of Cartesian-mechanistic science. These include areas such as astronomy, healing, agriculture, study of plants, animals and natural phenomena. Yet Indigenous science extends beyond these areas to also include a focus on spirituality, community, creativity, appropriate technology which sustains environments and other essential aspects of human life (Peat, 1996).[1]

Indigenous science includes exploration of basic questions such as the nature of language, thought and perception, the movement of time, the nature of human feeling, the nature of human knowing, the nature of proper human relationship to the cosmos and a host of other questions about natural reality. Indigenous science is the collective inheritance of human experience with the natural world. It is a map of reality drawn from the experiences of thousands of human generations which gave rise to a diversity of technologies for hunting, fishing, gathering, making art, building, communicating, visioning, healing and being.

There are those who would argue that there is no such thing as Indigenous science, that science is an invention of modern Western society and that Indigenous peoples have a body of cultural folklore, living practices and thought which cannot be considered a rational and ordered system of theory and investigation comparable to anything found in western science. Whether there exists an indigenous science in western terms is largely an incestuous argument of semantic definition. Using western orientations to measure the credence of non-western ways of knowing and being in the world has been applied historically to deny the reality of indigenous peoples. The fact is that Indigenous people <u>are</u>, they exist and do not need an exter-

81

nal measure to validate their existence in the world. Attempts to define indigenous science, which is by its nature alive, dynamic and ever changing through generations, fall short, as this science is a high-context inclusive system of knowledge.

Indigenous science offers both challenges and opportunities for math and science education since its insights and processing of knowledge parallel what many of the most innovative and reflective thinkers in education are advocating we do to extend our use of science to address the ever more pressing and complex problems we will face in the 21st century.

The development of simultaneous exploration and comparison of indigenous science and western science can provide the foundation for the flexibility and creative orientation to thinking and application in science that is essential for the future of human societies. **Scientists of aboriginal heritage though few in number, have the opportunity to become leading advocates for such a necessary re-thinking and transformation.** This presentation explores the nature of such re-thinking in terms of teacher preparation and teaching in science and math.

Western science is founded upon the premises of objectivity, abstraction, weighing and measuring. "If it cannot be tested, it does not exist!" is an often voiced credo of the mainstream scientist. Yet the deliberately limited focus on objectivity can block deeper insight into the metaphysics of the reality and process of the natural world. Western science does not consider the affective, intuitive and soulful nature of the world (Peat 1996).

In comparison, the Indigenous perspective is wholly inclusive and moves far beyond the boundaries of objective measurement. Indigenous science honors the essential importance of direct experience, interconnectedness, relationship, holism, quality and value. These and other qualities of Indigenous Science are largely unquantifiable, but are nevertheless honored as essential to gaining a deeper understanding of natural relationships. Such unquantifiable qualities are also associated with Western Science, as reflected in the Heisenburg Principle, where "the observer affects the observed". These unquantifiable qualities are usually not acknowledged in Scientific Findings, but clearly have an impact. The include the human factors, such as emotion, love, fear, disposition, personal desires and motivations to name but a few.

It is true, of course, that the Indigenous perspective is viewed as little more than primitive animism and sentimentalism by orthodox science. The Indigenous perspective from the view of orthodox science is an object for study but is anything but science. However the Indigenous perspective has the potential to give both great insight and guidance to the creation of the kind of environmental ethics and deep understanding which we must gain as we enter the critical times ahead. The serious study of environmental relationships to land, plants, animals and natural resources can provide much needed models for understanding the nature of environmental sustainability as a working process. Indigenous science engenders in its very process and content the revitalization of our human "biophilic" sensibilities, that is, our natural sense for affiliation with other living things. It also provides a way to reconnect with the soul which lies deep within each of us. As a system of thought and process of application it can provide an expansive paradigm for applying scientific understanding. Indeed, the perception of science must be expanded to make it the whole and comprehensive way of human knowledge it needs to become.

Indigenous consciousness defines itself in the experience of personality: the ego as agent, separate and simultaneously connected and previous to other egos, the land, seasonal cycles, spirits, and the world of transcendence, dreams and ancestry (Peat 1996).

What then are the philosophical orientations of Indigenous science? From an Indigenous perspective everything is considered with its own life and spirit which moves it through its field of relationships. All things are related in dynamic, interactive and reciprocal relationship. American Indian people were interested in finding the proper moral and ethical path upon which human beings should walk. In the search for this path knowledge was sought through individual and communal experience, environmental observation and information received from vision, ceremonies and spirits.

In the structure of the Tribal universe no body of knowledge exists for its own sake outside of a moral framework of understanding. Humans are co-creators with the higher powers of nature so that everything that we do has importance for the rest of the world. Also, everything that we experience has importance. All of our experience is a circle of learning, living and relationship. Education from this standpoint is totally inclusive of information from every source needed to make a decision in a moral and ethical relationship. All relationships have a history. People have a history of relationship with each other and with plants, animals, a land and the forces of Nature.

Indigenous culture is oriented to a place, a sacred bounded space. There is a historical and mythologically transmitted map of sacred sites. There are places in this bounded space where the People live. Learning is a matter of understanding, explaining and honoring the life the People are tied to in the greater microcosm. In terms of sacred sites, learning is a matter of remembering those natural forces which keep things in order and keep things going. Learning is a matter of respecting the understanding of how to use and become a contributing part of the place which supports the community by preserving and perpetuating its ecology. Indigenous science is tied to "place" and there is a profound connection that Indigenous people feel for their lands, to "that place that Indian people talk about," as it is referred to by Acoma Pueblo poet Simon Ortiz. For indeed, it is true that "in sacred place we dwell."

In defining itself, Indigenous science is internally consistent and self-validating. Its definition is based on its own merits, conceptual framework, practice and orientation. It is a disciplined process of coming to understanding and knowing. It has its own supporting metaphysics about the nature of reality. It deals with systems of relationship. It is concerned with the energies and processes within the universe. It provides its own basic schema and basis for action. It is fully integrated into the whole of life and being which means that it can not be separated into discrete disciplinary departments.

The processes of Indigenous science parallel and at times intersect with those of Western Science. Observation is emphasized. Indigenous people carefully observed aspects of Nature such as plants, animals, weather, celestial events, natural structures and ecologies of natural communities. They experimented with applications of their knowledge in the context of the environment or situation which was appropriate. There was no deliberate effort to decontext experimentation by moving beyond observation. In an Indigenous context prediction was not associated with the ability to control but primarily with gaining understanding of a natural process; for this reason science meant establishing relationships which led to establishing and maintaining harmony. Indigenous science intuits the desired results and then enters into specific relationships to accomplish its aims; it stresses direct subjective experience and close relationship to Nature.

An indigenous society stresses order and harmony but it also acknowledges and honors diversity. Relationships and renewable alliances take the place of fixed laws, and science accepts the possibility that chance and the unexpected can enter and disturb any scheme. Thus the circle is left open and chance as represented by the clown, the trickster and gambling games, occupies an important role.

Like Western science, Indigenous science has models which are highly contexted to tribal experience, representational and focused on higher order thinking and understanding. Models include symbols, numbers, geometric shapes, special objects, art forms, songs, stories, proverbs, metaphors, structures and the always present circle.

Finally, just as Western science uses physical tools to extend the range of its exploration of Nature, Indigenous traditions rely on the preparation of the mind and heart as well as physical tools. Indigenous science is as much a result of this preparation as it is a result of its accumulated base of knowledge. Therefore, if there is to be a true exchange of knowledge and mutual support the foundations of both systems must be appreciated as complementary ways of teaching, knowing, explaining and exploring the natural world. This why new approach to Native American science education which integrate old forms of education with new must be forged. The curriculum model presented in this work is one of the possible approaches.

PART THREE...
Footnotes

Chapter 7

1. For further information on the use of storytelling with a specific traditional Native American context, see Turpin, Thomas Herry 1975, <u>The Cheyenne Worldview as Reflected in the Stories of their Culture Heroes, Erect Horns and Sweet Medicine.</u> Ph.D. Dissertation, University of Southern California.

2. For further reference to the role of dreaming among Native American tribes refer to Garfield, Patricia 1974, <u>Creative Dreaming</u>. Simon and Schuster, New York, pp. 59 - 78.

3. Though the spiritual was the main focus of this learning, other learning which took place included the areas of social history of a tribe, human ecology and natural philosophy. For further information see Hultkrantz, 1967.

Chapter 8

1. "Mindset" as it is used in this context refers to a way of seeing and relating to one's environment within the parameters of a particular time and place. As a term, "mindset" is more inclusive than simply a perspective since it forms the basis for unconscious actions and attitudes to a particular situation or environment.

2. The case for drawing as an integral part of the process of creative visualization has been well developed in the work of Betty Edwards. For further information, see Edwards, B. (1986). <u>Drawing from the Artist Within.</u>

3. Steiner, V.J.(1986). <u>Notebooks of the Mind.</u>

Chapter 9

1. For further reference see: "Symbiotization of Cultural Heterogenuity: Scientific, Epistemological, and Aesthetic Bases" by Magorah Maruyama in <u>Anthropology for the Future</u> (1978).

2. For further reference see Brown (1982), Hultkrantz (1967), and Deloria (1973).

3. Resonance is herein meant to refer to the intuitive, cultural, and philosophical dimensions of relationships inherent in Native American orientation to nature.

4. The "seeking of life" is the key mission and philosophy behind all Pueblo ritualistic activity. This is a conceptualization which has taken various forms among all Indian tribes. If anything can be said to be held in

common by Indians throughout the Americas, it is this goal and philosophy.

Chapter 10

1. This reference: Peat, David. (1996). <u>Lighting the Seventh Fire</u>: contains an extended and insightful presentation of Indigenous science as seen through the eyes of a theoretical physicist. As such Peat's work represents the new openness on the part of some western scientist to consider the validity of Indigenous science and the Indigenous way of knowing. This chapter presents a summary of Peat's insights on indigenous science as viewed through my own life experience and indigenous viewpoint.

PART FOUR...
Science as a Cross Cultural Discipline

Chapter Eleven......The Native American Learner

Few studies have systematically explored the unique and culturally conditioned learning characteristics of Native Americans. Until the relatively recent interest in field sensitive vs. field independent orientations by some minority group learners, few researchers had focused upon the notion that the most effective way to educate was to develop teaching and learning strategies around distinct learning styles. Based on the concept of "cultural deprivation," the predominant notion has been to change the learning style through "educational reconditioning" so that students would conform to the mainstream educational system. From the earliest missionary attempts through the boarding school era to the present stage of public school education, Native American education has been dominated by attempts at "reconditioning" Native American learning styles.

Fortunately, with the introduction of self determination and the concurrent trend of cultural revitalization, this situation is beginning to change. In order to continue such a movement toward a more culturally relevant and learner sensitive education, some very important factors must be considered.

Significant learning is directly related to the degree of personal relevance the student perceives in the educational material being presented. The basis for such a premise stems from the idea that motivation toward any pursuit is energized by one's own constellation of personal and socio-cultural values. In the Native American social psyche this constellation of values has very ancient and well-developed roots. It is because of this embeddedness that Native American social personalities remain so durable and relatively visible through layers of acculturation. Understanding and utilizing this cultural constellation of values is a key to motivating learning in Native American education.

Since the turn of the century, Native Americans have experienced various levels of acculturation. Acculturation developed configurations of language and culture characteristic of the changes a particular Native American group underwent. For instance, many Native American students can be classified as being "English dominant," which has ramifications for teaching science. For while many are English dominant, they have been exposed, through home and community, to various levels of thought concerning the ways in which their particular tribal groups have traditionally viewed the natural world. There is often a real identification with both the cultural and linguistic revitalization of their particular cultural group. This sense of identification with tribal roots can provide a prime source of motivation to learn about science as it relates to an individual's heritage.

In addition to students rediscovering their tribal identities, there are students who are more completely bilingual and bicultural. These students generally want to continue to learn and live within the context of both cultures. Instruction in bicultural science for these students constitutes a real enrichment of their attitudes toward science and reaffirms cultural ties and identifications with their tribal groups. Bicultural science instruction for these students provides a means of bridging the sometimes great differences in mindset concerning natural phenomena.

Knowledge of core cultural values of Native Americans and an understanding of how such values differ from the values implied in American education is essential in bicultural education. The following example from Keresan Pueblo Indian philosophy may help to illustrate the origins of a particular set of Native American core values.

"Thinking Woman," a mythological being, plays an important role in the process of human valuing. Thinking Woman orders the universe by maintaining a balanced interrelationship between four worlds of being. The first world consists of collective prior human experience (similar to Jung's collective unconscious). The second world describes learning and the development of the individual. The third world carries the development of thinking, especially at its higher levels. The fourth world is a synthesis of all life within oneself, the individual life cycle.

All of these worlds are so intertwined that there is perpetual movement of our being in each of these worlds simultaneously.[1]

In contrast to the notion of some scholars that Native American cultures tend not to conceptualize abstractly, the Keresan philosophical concept of "Thinking Woman" is highly abstract. In fact, "Thinking Woman" as an abstract concept is characteristic of the traditional mindsets underlying the majority of Native American philosophies. Concepts like "Thinking Woman" have direct influence on traditional, spiritual and intellectual valuing.

Two interrelated valuing processes are involved with the concept of "Thinking Woman" and the four worlds of being. The first is called "Ma-shra," the immediate perceptions and the valuing therein based on the individual's experience of the immediate environment. The second is called "Shae-tah-ea" or "like this it is" and refers to learning by being shown, and the valuing resulting from such teaching.

"Thinking Woman" can be thought a process and frame of reference upon which core cultural values are formed. The goal is a balance of those things which are valuable to the life and harmony of the Keresan (Strum and Purly 1985).

Because of the increasing influences of the assimilation of American values by Pueblo Indians, this traditional framework for valuing is undergoing great change. All Native American cultures are experiencing similar transformations.

Let's look at individual, generalized categories reflecting personal value constellations: "the Pueblo Indian that is," "the Pueblo Indian in transition," "the new-found Pueblo Indian," and "the Pueblo Indian that isn't." "The Pueblo Indian that is" is someone who reflects values firmly embedded in the traditional Pueblo mindset. "The Pueblo Indian in transition" reflects the value sets of both the traditional culture and that of American society. The "new found Pueblo Indian" is usually an individual who has not been raised in a Pueblo context and is consciously in search of traditional roots. The value set of this group is externally and acutely oriented to an "idealized" standard of traditional Pueblo culture. The "Pueblo Indian who isn't," for a variety of environmental and personal reasons, has consciously decided to adopt mainstream American cultural values (Strum and Purly 1985).

In reality the above examples are highly dependent upon individual circumstances and can even alternate as dominant tendencies toward valuing in the same individual during different phases of his or her life. Human cultural valuing is a dynamic, evolving process, and human cognitive development does not fit neatly into categories. "Thinking Woman" and Pueblo valuing illustrate the variety of value sets characteristic of all Native American cultures and individuals in contemporary American society.

Pueblo Indians are among the most tenacious of Americans in the preservation of their traditional culture. Yet even among the Pueblos, the transition of values has a direct effect on their attitudes toward education. Core cultural values of Native Americans and their influences on attitudes and behaviors are relatively submerged since such values tend to operate at the subconscious level.

These values, however submerged, invariably affect the outcome of their educational pursuits. If the student can be made aware of cultural values of his/her people, learning will follow. Showing the student how what is being presented in a particular area such as science is relevant to or enhances the understanding of those cultural values will help him/her to learn.

The student's values play a key role as psychological "energizers" for the positive evolution of self-image. The accelerated rate of change since World War II has increased the inconsistencies in worldview and cognitive fabric of Native American life resulting in much intrapersonal tension. This conflict has given rise to a variety of emotional and social problems whose ramifications are poorly understood. But a subtle, well-integrated and consistent cognitive map and world-view is conducive to healthy concepts of self and positive social adjustment. The opposite is usually apparent when there are acute or chronic inconsistencies and conflict between the internal constellation of values and those of the external social environment.

Cultural content will facilitate educational goals and the development of students both intellectually and socially. Bringing core cultural values from the subconscious to the conscious sets the stage for the creative synthesis and interpretation of those values in a new and psychologically rewarding context.

The following selected, idealized Native American core cultural values underlie contrasts between traditional Native American and non-Native American values and associated behaviors and attitudes. These are not meant to reflect the wide variations within the Native American population related to levels of cultural assimilation nor the differences between Native American cultures. [2]

Personal Differences

Native Americans traditionally respected the individual differences in people. Not interfering in the affairs of another and verbalizing one's thoughts or opinions only when asked are common Native American expressions of this value. That others return this courtesy is expected by many Native Americans as an expression of mutual respect.

Quietness

Quietness and being still are values which serve many purposes in Indian life. Historically the cultivation of this value was survival oriented. In angry or uncomfortable social situations many Indians remain silent. This may be viewed as indifference by non-Indians, when in reality it is a very deeply embedded form of Indian interpersonal etiquette.

Patience

In Native American life the virtue of patience is based on the notion that all things unfold in their own time. Like silence, in earlier times patience was a survival virtue. In social situations patience revolves around respect for individuals, group consensus and "the second thought." Overt pressure on Indian students to make quick decisions or responses without deliberation should be avoided within most educational situations.

Open Work Ethic

In traditional Indian life, work revolves around a distinct purpose and is done when it needs to be done. The non-materialistic orientation of many Indians is directly reflected by this value. Only what is directly needed is accumulated through work, which is always tied to a specific job. In formal education, a rigid schedule of work for work's sake (busy work) needs to be avoided since it tends to move against the grain of this traditional value. School work must be shown to have a direct and immediate purpose.

Mutualism

In earlier times mutualism as expressed through cooperation was a basic survival value. As a value, attitude and behavior, mutualism permeates everything in the traditional Indian social fabric. "Solidarity" (group security and consensus) is highly valued. In American education the tendency has been to stress competition and work for personal gain over cooperation; the emphasis on grades, personal honors, etc., are examples. With Indian students this tendency must be modified by incorporating cooperative activities on an equal footing with competitive activities in the learning environment.

Non-Verbal Orientation

Traditionally, most Indians have tended to prefer listening rather than speaking. Talking for talking's sake is rarely practiced. Talk, like work, must have a purpose. Small talk and light conversation are not especially valued except among very close acquaintances. In Indian thought words have primordial power so that when there is a reason for their expression, it is done carefully. In social interaction, the emphasis is on the affective rather than verbal. This is an important consideration in lesson presentation and planning in that activities like a class discussion or too many questions should be avoided. It is because of this general characteristic that many Indian students feel more comfortable with the lecture/demonstration method. Inquiry approach, role playing or simulation are still valuable if presented with full understanding of this characteristic.

Seeing and Listening

In earlier times, hearing, observation and memory were important skills to develop since all aspects of Native American culture were transferred orally or through example. Storytelling, oratory, and experiential and observational learning were highly developed in Native American cultures. In an educational setting the use of lecture/demonstration, modified case study methods, storytelling and experiential activities can be highly effective if presented from a Native American perspective. A balance between teaching methods that emphasizes listening and observation as well as speaking is important.

Time Orientation

Time and its use have traditionally been considered relative. In the Indian world things happen when they are ready to happen. Time was relatively flexible and not structured into compartments as it is in modern society. Because the structure of time with its precise units is a hallmark of the traditional American curriculum, this conflict becomes a problem. The scheduling of activities, fixed class hours and measured time allotment for lesson presentation of Western curricula can precipitate disharmony between the traditional Indian learner and the material being presented. The solution is to allow for flexibility and openness in terms of time within practical limits.

Traditionally, Indians oriented themselves to the present and the immediate tasks at hand. This orientation stems from the deeply embedded philosophical emphasis on Being rather than on Becoming. Present needs and desires tend to take precedence over vague future rewards. Although this orientation has changed considerably over the last forty years, vestiges are apparent in the personality matrix of many Native Americans. Learning material presented should have relevancy for the time and place of each student.

Practicality

Indians have the tendency to be practical. Many Indians have less difficulty comprehending educational materials and approaches which are concrete or experiential than those which are abstract and theoretical. Learning and teaching should begin with numerous concrete examples and activities to be followed by discussion of the abstraction.

Holistic Orientation

Indian culture, like all primal cultures, has a very long and well-integrated orientation toward the whole. This is readily apparent in various aspects of Indian culture, ranging from healing to social organization. Presentation of educational material from a holistic perspective becomes an essential and very natural strategy for teaching Indian people.

Spirituality

Religious thought and action are integrated into every aspect of the fabric of traditional Native American life. Spirituality is a natural component of everything. As a general educational tool, this presents a good advance organizer for concept presentation in that all aspects of Indian culture are touched by it. Discussion of the general aspects of spirituality and religion is an important part of the curriculum, although care must be taken to respect the integrity and sacred value of each tribe's religious practices and inherent privacy. Ideally, all discussions of Native American religion should be kept as general and non-specific as possible. Specifics should be discussed only in the proper context and with the necessary permission of that particular tribe.

Caution

The tendency toward caution in unfamiliar personal encounters and situations has given rise to the stereotypical portrayal of the "stoic Indian." This characteristic is closely related to the "placidity" and quiet behavior of many Indian people. Such caution results from a basic fear regarding how thoughts and behavior will be accepted by others with whom they are unfamiliar, or in a new situation with which they have no experience. Within the educational context, every effort should be made to alleviate these fears and show that the student's subjective orientations are accepted by the teacher. The class and lesson presentation should be made as informal and open as possible. An open friendliness and sincerity are key factors in easing these tensions.

Classroom Discipline

Among most Indian people, the cultivation of self-discipline is valued. Behavior is regulated through group and peer pressure. Withdrawal of approval, shame and reflecting unacceptable behavior back to the individual are the main forms of punishment in the traditional Indian context. In the classroom, direct and demeaning personal criticism in front of others is considered rude and disrespectful and can lead to "loss of face" and complete withdrawal and alienation by the student. A clear understanding of the consequences of behavior is a must for teachers.

Field Sensitive Orientation *(Group orientation, social relationships)*

A significant number of Native Americans have a tendency to express field-sensitive behaviors as opposed to field-independent behavioral characteristics. This tendency has direct implications for the learning styles Native Americans tend to exhibit.

PART FOUR...

Chapter Twelve......Border Crossings

The understanding of science as a subculture of Euro-American society has begun to be applied at some levels of main stream science education. This understanding has also energized further research and implementation of cross-cultural science curricula tailored to the needs of Native American learners.

The learning of Western science by Native American students has always required various levels of cultural adaptation. Such cultural adaptation can be said to be akin to the "crossing of cultural borders". That is, Native American students cross cultural borders, from the familiar contexts of peers, family and tribe, to school, school science and the actual sub-culture of science. The notion of border crossings implies that students do not leave their home culture behind when they enter the cultural landscape called "school science." Rather they become explorers on a mission to learn about a "new territory" to gain knowledge and understanding that they may use back home toward their own self determined and practical ends. These practical ends include preparing for a career, economic development, environmental responsibility and cultural survival at the community level (Aikenhead 1997:1).[1]

However, in interactions between Indigenous cultures and the subculture of Western science, profound conflicts arise. Their orientations differ in terms of: survival vs. power over nature and other people; coexistence with the mystery of nature vs. attempting to explain the mystery of nature away; the search for intimate relationship with nature vs. decontextualized objectivity; and accommodation, intuitive and spiritual vs. reductionist, manipulative and analytical. (Ibid p. 6)

In addition, "Indigenous knowledge of nature tends to be thematic, survival oriented, holistic, empirical, rational, contextualized, specific, communal, ideological, spiritual, inclusive, cooperative, coexistent, personal, and peaceful." (Ibid p.7).

This essential orientation difference challenges Native American students as they attempt to cross the borders into the subculture of Western science as represented in schools. If the teaching and learning of science is supportive of the student's culture orientation, "enculturation" is the result. If the teaching and learning of science is at odds with the student's cultural orientation, the result is "assimilation" forcing students to abandon or marginalize their way of knowing to reconstruct a new (generally dysfunctional) way of knowing. Unfortunately, the latter is more often the case (Ibid p.10).

The essential question is : How can students from Indigenous cultures learn non-Native subjects like science without being assimilated harmfully by the underlying value structure?

As Native American students grow up they intuitively develop a facility to cross the everyday worlds of peers, family and community into the sub-cultures of schools. This natural tendency of students' negotiating cultural borders can also be applied to the learning of school science. In facilitating this kind of cultural border crossing students and teachers interact in a creative expression of cultural adaptation. In this creative expression students act as explorers and teachers as guides in a metaphoric journey through the cultural landscape of Western science (Pomeroy 1994:13-50).

Yet, the reality of student-teacher interaction with regard to science learning is wrought with difficulty. Negotiating meaning from one domain of meaning to another can be complicated. Students generally get very little help doing this kind of border crossing. Few teachers are inclined to assist students, and if they are, they have few resources for being trained in this kind of cross-cultural negotiation. (Hennessey 1993: 9).

For example, (Phelan 1991) identified four worlds for student transitions. These include: a congruent world which supports smooth transitions, a different world which requires transitions to be managed, diverse worlds which lead to hazardous transitions, and highly discordant worlds which cause students to resist transitions and in which they become virtually impossible (Phelan 1991).

Following a similar line of research, Costa (1995) divided students in science classrooms into five basic categories:
1. "Potential Scientists" cross borders into school science so smoothly and naturally that the borders appear invisible;
2. "Other Smart Kids" manage their border crossing so well that few

express science as being a foreign subculture;
3. "I Don't Know Students" confront hazardous border crossings but learn to cope and survive;
4. "Outsiders" tend to be alienated from school so the border crossing in school science is virtually impossible; and
5. "Inside Outsiders" find the border crossing almost impossible because of overt discrimination within the school. (See Aikenhead1997)

Assisting students to develop the skills for accessing Western science for achieving goals defined by a "philosophy" of Native American science education must be a key aim in the development of science curriculum for Native American students. Determining what kinds of skills and knowledge are appropriate for Native American students to learn with reference to economic development, environmental responsibility and cultural survival is the next step of developing such a comprehensive process. Sound integrated education that helps students be flexible and adaptable and enhances their ability to train on the job presents the most strategic form of science education for these students. An approach which integrates scientific, technological and Indigenous knowledge into real life situations and issues has the best chance of being effective. Participatory research is one way of accomplishing this.

Jenkins (1992) argues that using science in everyday situations requires changing knowledge into new forms which can be applied to problems and issues at hand.

Restructuring scientific knowledge into new forms for Native contexts requires knowledge of both a different cultural orientation and a different approach to teaching and learning science. **Essentially Native knowledge comes already contexted and ready for use, Western scientific knowledge does not**. As this is the way Western science is taught in school, it is no wonder that may students cope by developing a view of science as apart from their real lives.

The integration of selected science and technology content in an Indigenous worldview requires coordination with relevant economic, social and resource needs of indigenous students and communities. A cross-cultural Science -Technology- Society (STS) model which has been used by science educators in third world countries is one possibility for facilitating this sort of integration. STS is a dedicated student oriented, critical and environmentally responsible approach to science, and it decontextualizes

Western science in the social and technological settings relevant to students (Bingle and Gaskell 1994; Akinhead 1997).

Applying an anthropological approach from an Indigenous perspective to the teaching and learning of Western science is another possibility since this promotes "autonomous acculturation,...(or) intercultural borrowing or adaptation of attractive content or aspects." This would be a more constructive and culturally affirming alternative for Native students than assimilating, or enculturating themselves to Western science. Students may act as anthropologists learning about another culture. Like cultural anthropologists they would not need to accept the cultural ways of their "subjects" in order to understand or engage in some of those ways (Akinhead 1997: 26).

Combining the STS approach with that of "**the student as anthropologist**" in the context of an indigenous perspective and community reality can form an ideal foundation for indigenous students learning science. The teacher's role is to learn to act as kind of **cultural broker** who assists students in handling cultural negotiation and conflict between views. Essentially, students act as "cultural tourists" in a constructive way and teachers take on the role of "tour guides" and "travel agents" as they help students cross the cultural knowledge borders between science and their own worlds.

The development of such a curricular approach can further be facilitated by studying the students' community reality and using that as a foundation for relevant and meaningful themes, then comparing that foundation with the subculture of Western science. It begins with the observation that Western science education is often most at odds with the diversity of socio-cultural environments from which students come. For example: Learning to hunt in Native American society is a programmed sequence of observations and experiences tied to a process which might include :
1) Learning the habits of the animal hunted (mythology, listening and observation);
2) Learning to track, read appropriate signs and stalk the animal (observation, intuition and reasoning);
3) Learning the appropriate respect and ritual to be extended to the animal hunted (learning a mindset);
4) Learning to properly care for the carcass of the animal once it has been taken (an ecological ethic, technology); and

5) Learning to fully utilize the various parts of the animal taken (technology).

These processes require teaching techniques ranging from formal instruction to experiential learning. All of these teaching/learning situations are directly related within a particular contextual framework necessary for conveying these forms of knowledge. Learning is directly tied to the job to be completed. It involves teaching to accomplish a specific goal. One observes and learns from that which one seeks to do. The teachers and situations are many.

Native American cultural education evolves around the problem of **learning how to do something**. By contrast modern Western education evolves around frames of reference which prepare students for possible future needs and tasks deemed important in a modern industrial and technological societal complex.

Within most typical American educational situations, what is learned is laid out in a distinct linear pattern. All that is to be learned is hierarchically mapped beginning with objectives to be reached in each grade level and moving to more specific units and individual lesson plans, each of which has objectives and associated learning activities. This highly structured and programmed approach is designed for easier teaching of large numbers of students and for consistency in what is learned. Yet if one views this approach in terms of addressing individual student learning styles, many problems become apparent.

Much of modern education imposes a preconceived psychological pattern of the "right and wrong ways to do things." This pattern imposes societal will on all those who participate in American public education. In the process many students are denied use of their innate repertory of intelligences and cultural styles of learning. Ability to learn by simply doing, experiencing and making connections will be significantly diminished through such a homogenization of the educational process.

PART FOUR...
Footnotes

Chapter 11

1. This conceptualization of "Thinking Woman" in Keresan Indian philosophy was related by Mr. Ted Purley of Laguna Pueblo to Dr. Fred Strum of the University of New Mexico. This segment was paraphrased from an article by Dr. Strum entitled "Pueblo Valuing in Transition," in the **Pueblo Cultural Center Newsletter,**" Albuquerque, NM, Spring 1985. Keresan-speaking Pueblos in New Mexico include Cochiti, Santo Domingo, San Felipe, Santa Ana, Laguna, Zia and Acoma.

2. These characteristics have been quoted or otherwise adapted from the **American Indian Education Handbook**, American Indian Education Unit, California Department of Education, 1982.

Chapter 12

1. A key reference which proposes the notion of "Border Crossings" is: Aikenhead, Glen S. "Toward a First Nations Cross-Cultural Science and Technology Curriculum" in <u>Science Education.</u> Vol. 81, 1997. This chapter summarizes the key elements of the concept of "Border Crossings" as they relate to the thesis and model presented in this work.

PART FIVE...
The Creative Process in Science and the Native American Learner

Chapter Thirteen......The Curriculum Design as Metaphor

The development and implementation of a concept in educational curricula is not very different from the work of the designer. To begin with, every good design must not only be technically competent, but it must also have a balanced and well-developed aesthetic sense. The design must reflect an understanding of the users and their needs. The designer must be able to appreciate and apply the thought processes and techniques of both the artist and scientist, blending and manipulating ideas and structures into a coherent product. Designers generally focus their attention on producing the desired results using solution-oriented strategies (Lawson 1980: 5-7).

The design of the curriculum presented in this work reflects the author's attempt to provide an exemplary model for an approach to science education aimed at Native American students that is both technically workable and aesthetically balanced. The curriculum design reflects a basic understanding of the socio-cultural characteristics of Native American students. It is a reflection of the thought processes and creative sensibilities of both the artist and scientist, integrated through the eyes of the educator. The curriculum design is the direct result of the manipulation of ideas, approaches and intuitions in the search of a "workable solution" to the alienation expressed by many Native American students to current standardized approaches to science education.

Two very simple ideas were utilized to guide the designing process of this curriculum: first, that culture is intimately involved in the nature and expression of the scientific thought process, and second, that science is a creative process of thought and action and, as such, is highly interrelated with other cultural systems such as art. Based on this twofold schema, a curriculum approach has evolved that presents science to Native American students from a cultural perspective and as a creative system of thought and action for understanding relationships between natural phenomena.

Every curriculum comes to life through its processing. Because of this, it is important to know the underlying strategy which guides the processing of a curriculum. All too often in modern education, strategy is ignored or poorly understood by those who implement that curriculum. The relative success of any curriculum approach is highly dependent upon knowing its underlying process strategy.

This section will show the curriculum strategy as it has been approached by this writer at the Institute of American Indian Arts. The analogy of a **mandala** and its concentric rings and that of **tracking** will be "worked" in such a way as to present a verbal and visual image, so the reader will have a sense of the dynamic process inherent in the experienced curriculum.

The symbol of concentric circles in world culture seems always to connote the process of an event. That is, the concentric ring when it is used in primal myth, ritual, or art, denotes that something happened here, or that something is happening here, a distinct natural phenomena or an important life activity, perhaps at a waterhole, or during a ceremony. The symbol images the fact that everything is unique and leaves a signature track, but that all share likenesses to be found in the overlap of the rings.

In the mythology, ritual, and art of the Australian aborigine, the concentric ring marks the place of an event of sacred significance and great insight. The mandala and the medicine wheel are other symbolic exemplifications of significant process events. Since myth mirrors and analogizes nature and all its relationships, it is no wonder that one of the simplest symbols represents one of the most complex processes of nature — that of interrelationship.

In this section, a **mandala** (relationship circle) shows the relationship and significance of the courses of study in the curriculum. Mandalas are mechanisms for focus and meditation and as such are open to interpretation depending on where one's mind/heart happens to be in space and time; this is true for both individuals and cultures. This mandala does not necessarily represent any specific Native American tribe and is predicated on the idea that everything has its proper place in continual and dynamic interrelationship with other entities in a natural environment. This interrelationship is also true for ideas, concepts and courses of study.

The mandala is meant to show graphically the way in which science is symbolically related to art, philosophy, psychology, creativity, healing, myth, religion and the natural world in Native American cultures. Science, like art, has never been viewed as a self-contained, classifiable "subject" in traditional Native American cultural contexts. Rather it has always been integrated to some degree or another with every other aspect of tribal life. (See Tracking a Symbol diagram on page 111)

When one focuses on one system, he or she necessarily becomes involved with the others. Usually, the designation of science, art or religion simply indicates the *degree* to which the characteristics, end products, aims or purposes of each system is emphasized over the others. Both primary and secondary thought processes, channeled in a culturally specific way, are utilized in the structuring of these systems. These, together with imagery and intention, form the centering point (nexus) for the nature and dynamic expression of creative thought in art, science and religion at the individual and cultural level.

At the beginning of the human quest for knowledge, there was a convergence of science and religion. This convergence continues to be the rule, rather than the exception in most cultural traditions of scientific and environmental knowledge. Indeed, until the Middle Ages, science was intimately interwoven with religion in Western societies as well. With the Scientific Revolution in Europe during the 16th and 17th centuries, culminating in the Newtonian-Cartian Paradigm, science was inexorably divorced from religion. Yet, even with the advent of Newtonian Science, the Judeo-Christian theology of Middle Age Europe provided the foundational ethos of man's dominion over nature. This concept of "the new science and the accompanying view of reality, of man and his place in the Cosmos and the purpose of nature and knowledge, led, as if with inexorable necessity, to the complete mechanization of the world and the accompanying de-spiritualization of both nature and human beings" (Ravindra 1991: 23).

Now, as we enter the 21st century and have a retrospective view after three centuries of Newtonian science and technology (and especially its fantastic growth and influence in the 20th century), there is a sobering realization that it can not be the sole basis for a sustainable world view. The parallel and evolving attitude is that older "Indigenous" perspectives on nature, life and spirit contribute something of value to human life and ecological sustainability. Indeed, with the evolution of Quantum Physics, a view of physical reality is being revealed which reinforces some of the central orientations of Indigenous Science and spiritual traditions.

The Quantum Science of the 21st century will no doubt cross paths with those ancient traditions, which in their own way, anticipated the scientific discoveries of the "new physics".

"It is no exaggeration to say that the future course of history depends on the decision of this generation as to the relations between religion and science", Ravindra 1991, 3).

The convergence of science and religion in the beginning of the journey for knowledge and again at the end, when we contemplate the "unknowable" is no accident, but rather the evolution of a more mature consciousness of the inherent interrelationship and interdependence of all aspects of nature. Indeed, as the Lakota saying reminds us, "Miki Wiyasin" (we are all related!)

The mythic symbol of Grandmother Spider Woman and her web provides a metaphoric analogy for explaining the symbol. In some Native American cultures, Grandmother Spider is the archetype of the all-knowing universal mother. Her web represents the inner and outer manifestations of nature as spokes radiate out of their essential cores. Because of her central position and the construction of her web, she is in constant communion with all that touches her web. She is an archetype of holism.

Alice Marriot and Carol Rachlin, in their book <u>Indian Mythology</u>, state the following:
"Her entire web is infused with <u>Power</u>, the animating force of the universe which derives from the Creator and his helpers. There is no other English word which even partially conveys what most Indians mean by 'Power.' Perhaps 'talent' or 'genius,' when the word is applied to an individual, most nearly approaches the meaning Indians give to the word 'Power' (Marriot and Rachlin 1968: 32).

The concentric ring is also a basic symbol of wholeness. The mapping of concentric rings of relationship is a major activity in primal peoples' mythology, ritual and adaptation to their respective natural environments.

In all of these rings of wholeness there is awareness of a particular aspect of nature, re-ordering and then representing that aspect in some form. This process is one of the universals of the creative act and as such is a primary dimension of science and art.

The rings are those which comprise observable interrelationships in nature. Every process in nature occurs in what can be called a context of concentric rings. The concentric ring provides a visual symbol of relationship, a way of visualizing how all processes affect other rings of other processes.

The pattern of thought in the mandala associated with this curriculum represents an orientation for the study of scientific thought and process. The descriptions and explanations are a representation of ancient ways man used to derive meaning in his relationship to the whole of nature. This mandala represents a map of mindsets relevant to traditional Native American contexts. Each of the circles represents a rich source from which to originate discussion of science based on metaphoric representations of perceptions of the natural world.

If the mandala shows how knowledge can grow and develop outward in concentric rings, concentric rings can also form the basis of learning how a learner can track ideas and intuitions, observe fields of knowledge and see patterns and connections in thought and natural reality.

Tracking involves good observation, common/natural sense, following an intuitive yet discernible direction and developing intuition and visual thinking. In the literal sense it simply involves observing the rings that are coming into you and quieting the rings that are going out from you.[1]

Tracks can be read from many perspectives. In reality, tracking strategy begins with scanning the rings of a landscape with a kind of macrovision. Such scanning eventually leads one through smaller concentric rings down to a micro-focus on a specific animal.

From this perspective tracking is intimately involved with learning how to see connections between rings. The analogy of tracking, then, can be used to illustrate an essential process in the learning of science, that of seeing connections, being aware of circles of interrelationship and **following the tracks of a problem or natural process**. The process of tracking is itself comprised of a group of concentric rings beginning with the physical, followed by the psychological, then the social and metaphysical. These rings of tracking represent interrelated dimensions of process and "field" thinking.

Field thinking within the context of tracking simply means becoming aware of a particular field of relationships and being able to pick out specific possibilities within this field which directly relate to what one wants to find or to do.

Tracking requires the ability to see connections of a physical nature. For example, an older hunter of wide experience in a particular environment can tell a fox is coming when a bluejay begins to scold in a certain way. Some time in the past the old hunter observed and heard a bluejay scolding a fox in just this way and fixed that image and sound in his memory. He saw a specific connection within a field of possibilities. When that particular bluejay scold is heard again, the hunter remembers the sound and the image.

At the physical level tracking requires the development of the ability to discern patterns using visual acuity, to discern differences in sound, feeling, smell and even taste. It involves the ability to know using these basic human perceptual abilities. Tracking is common/natural sense because it requires using and developing these latent abilities of the whole mind and body. It exercises visualization, intuition and reason coupled with a knowledge of an animal gained through experience and observation. In this sense, it is a unification of the artist and scientist. Herein lies the connection between the actual physical process of tracking and the process of scientific investigation.

The creative scientist is like Sherlock Holmes in that he is an intellectual tracker. He observes an intellectual track just as an animal tracker observes a physical track. The creative scientist then asks questions. He gets down and looks for clues, for patterns. He uses his common/natural sense. He uses his intuition, his logic and his capacity for visualization. If he has an idea, a hunch, a concept, then he searches for evidence that it really works. Just as in tracking an animal, finding the first track (first insight) is relatively easy. Finding the second and third track, and finally the trail and solution, is what requires the application of one's unique intellectual and intuitive abilities. Science involves experiencing the world, seeing, then tracking, then representing some of nature's concentric rings in a particular form. Making ties through these process relationships is the essence of "science-in-the-making." Science is practice in working these ties to develop a more complete perspective of a particular problem.

Tracking in the metaphysical sense is basically following the concentric rings of the physical, psychological, social and spiritual to their various origins. For example, in primal hunting and animal mythologies one visualizes tracking the trace of an animal in the eye of the hunter, then into the mouth of the hunter, then back through his hand, his body, and his psyche in the forms of art, dance, song, and ritual. Through myth and its associated rings of expression the hunter celebrates the animal to make more animals, to dance more animals, to increase the fertility and vitality of certain animal species, and in doing so to keep the concentric rings rotating and interrelating in a positive way.

Primal mythologies abound with examples of tracking and working the tracks of the ancestors through time and through a geographical landscape of mythology whose concentric rings radiate to the present time and place. A key to understanding mythological tracking of concentric rings is developing the ability to think "upside down."

In mythological contexts, things are reversed and inside out. For instance, the peyote hunt of the Huichol Indians of Mexico is characterized by tracing of the steps of the Huichol ancestors to the mythological land of Peyote, which is called "Wirikuta." This reverse tracking of the ancestor's steps occurs over a geographical and mythological landscape in which those Huichol seeking the peyote are led by a "Mara'a kame" (Huichol Shaman) through five concentric rings of relationship. Each of these rings is symbolized by its own state of mind, its own ritual, its own natural energy and geographical landmark. The geographical landscape from the Sierra Madre where the Huichols now live to the desert flats just outside of San Luis Potessi where the peyote cactus is to be found represents the trail of the ancestors' tracks. Along the geographical landscape of this trail are natural landmarks which are representative of the concentric rings of important natural and life sustaining energies of the earth. These are the archetypal energies of earth, wind, fire, water, plant and animal. Each of these energies are represented in Huichol traditional yarn painting by mythological animals, beings and entities. They are symbolic of the natural shaping energies of the earth's landscape.[2]

Tracking the ancestor's mythological steps, then, takes the Huichol through different levels of knowing in reference to peyote, Huichol origins and myth, Huichol cultural philosophy and the natural energies of the Huichol's natural habitat. This sort of metaphysical tracking through concentric rings of interrelationships illustrates how landscape, natural energies, plants and animals affect each other. Noting these relationships and their mutual effects is the beginning of "primal" science.

For instance, within the contexts of Native American mythologies, certain geographical features personify ties between natural processes. Generally, such features are looked upon as sacred places. These natural features may be specific formations, springs, lakes, rivers, mountains or natural places. All these features, physically, visually and metaphysically represent concentric rings in nature. Many are symbols of life sustainers such as corn, deer, buffalo, fish, rain clouds and forests. An understanding of the relationships inherent in these ties is essential to survival. Therefore, much attention is given to ways of knowing and learning about important natural phenomena.

Myths present a way of mapping a particular geographical landscape. Relating the stories associated with a particular geographic place is a way to begin to develop a cognitive map of that place and of its concentric rings of interrelationship. Migration myths, for instance, are tracking stories through a geographical landscape. In many Native American migration myths it is implied that the ancestors left representations of themselves in various natural forms or phenomena to remind people how to act and how to relate to the natural world.

Through the symbol of concentric rings, myth is able to give us a visual image of how one thing in reality is like something in myth. Every myth has its concentric rings of meaning and is told and retold in this way. The telling of a myth begins with a simple version for children, then moves to a slightly more complicated version for adolescents, then to a deeper version for initiates, and then to a still deeper version for the fully mature.

Science in process follows tracks in a particular field or level of natural reality. This tracking requires opening one's mind to the possibilities within each of the many concentric circles within that dimension. Learning to blend the mythological, aesthetic, intuitive and visual perspectives of nature with the scientific, rational and verbal perspective is an integral part of this curriculum model. Science education, from this viewpoint, involves learning to see nature holistically. This requires a continual shifting and interplay between the two complementary perspectives of the scientific and the metaphoric. Facilitating the orchestration of these two ways of viewing nature is an intrinsic goal of the curriculum model.

Finally, to illustrate one of the many of possibilities stemming from this strategy, I present one last example: In this curriculum approach, the track is a symbol. Visual and metaphoric symbols abound in nature and Native American mythologies. These symbols are the connection or keys which access the myth and the relationships of concentric circles, and knowledge of natural realities. In teaching and learning a process discipline such as science, beginning with a mythological track and following that track through its concentric circles from its abstraction to its reality and then back again presents a natural and creative approach.

The Southwestern Indian symbol of the humpback flute player, sometimes called "Kokopeli" or "antman," is a mythological symbol which represents the notion of seeds, fertility, sexuality, abundance and the spread of art and culture. The Kokopeli is a natural process symbol which is pregnant with meaning. [See Diagram: Tracking a Symbol through Concentric Rings of Relationships]

As such, the symbol of Kokopeli is surrounded by many myths and metaphors representing various dimensions of the procreative processes. Each of these processes is encircled by a body of aesthetic and cultural expressions tied to physical realities observable from a scientific perspective.

This strategy for gaining perspective begins with exploring the creative dimensions and potential of this symbol to exercise visual and verbal modes of thinking and representation. This then leads to conceptualization followed by an exploration of derived concepts which leads one directly to scientific investigation. Intertwined with these rings are those of the student: his/her personality, his/her style, socio-cultural environment, the school environment, and the process of science learning. Around these are the concentric rings of the teacher, his/her personality, his/her style of teaching, socio-cultural environment, school environment and the nature of the science curriculum. All these rings affect each other and form tracks of process which are essential considerations for relevant and effective science learning.

> Tracking is a "process art."
> Learning is a "growth process."
> Teaching is a "facilitating process."
> Life and Nature are always "relationships in process."

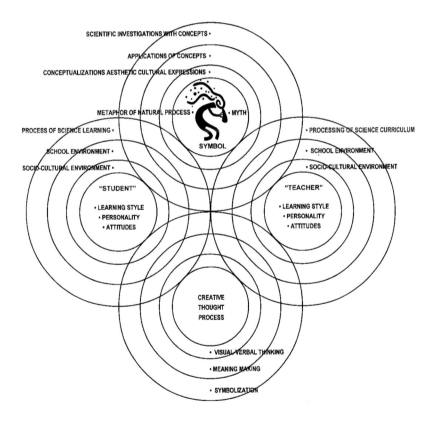

Teaching is a communicative art which is implicitly or explicitly involved in the transmitting of some level of these cultural symbol systems. The scientific thought process, or "logical-rational-thematical" level of thinking is, like other thought processes, culturally conditioned in practically all phases of its development. The way Native Americans characteristically communicate about nature or symbolically represent natural processes provides a key frame of reference for this approach to science education.

Art presents a natural vehicle for the presentation and elaboration of science content from a sociocultural and affective/creative perspective. For a scientist, such learning activities revolve around basic research and the context of experimentation. For an artist, these learning activities revolve around the creative evolution of a piece of art. As Herbert Read so aptly stated, "Art is the representation and science the explanation of the same reality" (Read 1945). This is precisely the Native American perspective.

In Native American cultures, science and art are complementary dimensions of the community mind. This subtle yet profound relationship becomes apparent only when one focuses upon the processes of thought as opposed to its end products. **It is, therefore, my contention that one can use art to teach science, and science to teach art, and cultural philosophy to teach both.**

Based on these foundations, the strategy of the curriculum model is to provide the presentation of the basic principles of general science by first introducing students to the ways in which these principles are communicated, utilized, or otherwise exemplified in Native American culture. Students are then presented with a comparison of these cultural examples with similar elements in Western science. The idea is to illustrate that these principles are the result of the creative thought process and to establish this as a point of commonality between both cultural perspectives. Finally, the students are provided with a variety of opportunities to review and apply the basic principles. It is not the purpose of this model to supplant the teaching of basic science principles through more conventional science curricula, but rather to facilitate their transfer through culturally-meaningful communication. [See Diagram for the Left Brain]

Selected Native American paradigms are presented reflecting aspects of nature, followed by group exploration and discussion using appropriate teaching/learning strategies based on the characteristics of the students and the particular requirements of the learning context.

Western scientific paradigms which correspond to, or differ from, those Native American paradigms, are introduced. An exploration and discussion is undertaken by the group using appropriate teaching/learning strategies. The group compares the Native American and Western paradigms. A partial synthesis and summary of the perceptions, paradigms and components of the selected images of Native American and Western sciences concludes this phase.

Phase 1

This first phase is to help the student form a frame of reference, or orientation, to the underlying patterns of thought which guide the expression of science in Native American and Western cultures. Becoming comfortable with what is being presented and understanding and internalizing some of the premises is essential to the development of thinking and to the exercising of creative capacities in the later phases of the learning process. In this first phase, the student ideally establishes connections and begins to see interrelationships and relevancy in what is being presented.

Phase 2

The second phase of the curriculum focuses upon specific principles, forms and classification structures of selected segments of Native American and Western cultural science. This phase begins with the general presentation of selected aspects of a particular Native American ethnoscience, forms of symbolic expression, or principles which relate to a specific group of natural phenomena. An analysis through appropriate individual or group teaching/learning activities follows. An attempt is made at describing the nature of the classification patterns that have been revealed. This process is repeated for selected aspects of Western scientific disciplines, forms of symbolic expression or principles. Finally, the group reviews and compares Native American and Western cultural content.

This phase extends and improves the students' understanding of the content being presented. It also involves practice in applying the "patterns" which have been discovered in phase one. The focus here is on specifics and upon the accumulation and exploration of examples which fit the patterns. This necessarily involves a great deal of reading and individual research. In short, this phase involves the further elaboration, processing and augmentation of the ideas presented and explored in phase one.

Phase 3

The third phase begins with the exploration of Native American symbolic orientations, forms or systems of creative processing as they relate to a specific area of ethnoscience. The focus here is on discovering the ways in which Native Americans' creative thought processes and problem solving affected the development and expression of ways they perceived and adapted to nature, and then on corresponding Western scientific symbolic orientations and processes. Finally, bridges of understanding between Native American and Western scientific approaches are constructed.

This phase revolves around integrating each system, as well as coming to terms with the differentness of each approach to science. It engages the creative capacities of each student through the use of relational, qualitative, interactionary and mutualistic patterns of thinking, bringing together opposites, acknowledging and understanding the interdependence of the two systems. This phase also involves the destructuring and then restructuring of what has been learned previously by the student. The student begins to explore and discover his or her own creative abilities and potentials. The ideal final goal of this curriculum is to produce a Native American student who is basically science literate, who understands and appreciates both Native American and Western approaches to science, and who has, in the process, become aware of his own creative thinking abilities.

The overall process is meant to parallel closely real life creative problem solving. First, a situation is perceived; it is analyzed in order to understand it and discover the problem; what is learned is then synthesized and new ideas are formed. These ideas are evaluated in search of the best ideas or solution; the best idea is applied and then reacted to.

This cyclic process of looking at a problem creatively in reference to one's cultural mindset is the most important aspect of the approach. It not only allows for a greater perspective and understanding of science, but it is transferrable to all other areas of the educative process.

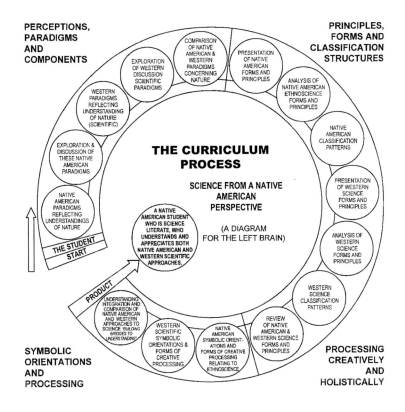

SCIENCE FROM A NATIVE AMERICAN PERSPECTIVE
A PROCESS ORIENTED STRATEGY

THE COMPONENTS OF THE CURRICULUM STUDY
(A DIAGRAM FOR THE RIGHT BRAIN)

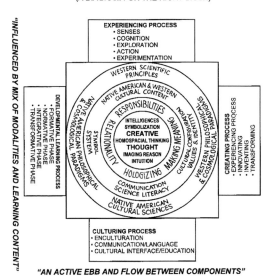

"AN ACTIVE EBB AND FLOW BETWEEN COMPONENTS"

PART FIVE...

Chapter Fourteen......The Courses

From the Native American perspective, science is an abstract, symbolic and metaphoric way of perceiving and understanding the world. From the Western cultural perspective, science is essentially practiced as a rational way to solve problems.

These two approaches can complement one another. Like the sacred twins in Native American mythology they are by nature intimately interrelated. Each derives its meaning from the other. Science as a whole is based on both the intuitive and rational mind. This curriculum represents one way of enhancing and flowing with this natural relationship, which Einstein expressed well in the maxim:

"The intuitive mind is a sacred gift,
the rational mind is a faithful servant."

The creative process is the most essential universal that centers people and learning and understanding. It is the elemental process in the natural as well as the world of thought. How to go about conditioning oneself for creative thought is what I try to establish as the first foundation in this science curriculum. The students I taught at IAIA for over twenty-five years were immersed in a variety of experiences and encounters which enlivened their ability to think creatively and to think in multidimensional ways using not only the five senses but all their faculties.

With students studying to become artists, the first step into the world of science learning had to derive from their own creative center. The next step was to evolve tools to help them as they undertook this journey. Learning through this curriculum process is a journey in which students make stops at each of these orientations.

There are many concepts and metaphors held in common by American Indians. One is the idea that each tribal group orients itself to its land and understands its land intimately. Most American Indian groups recognize seven basic orientations: the four cardinal directions, north, south, east, west, the center, usually the community, itself or village, or the center of the territory, and then the Above and the Below, the below representing the earth and the above, the celestial or universe.

The following description adapts this Native American construct of orientation to the purposes of envisioning a metaphoric curriculum model for creatively teaching science to Native American students, [See Curriculum Mandala Diagram]

First is the "centering place," the womb, which contains the essence of all that emanates from it. In Native American mythological terms, it is the primal source of everything, symbolized by balance and by all the colors of the rainbow, the dwelling place of the spirit and the essence of creation. In this curriculum, it is the domain of the creative thought process and holistic thinking.

The first direction, East, the domain of the rising sun, is the source of "First Light," dawn, purification and insight. Its characteristic process and product is symbolized by Wisdom and by the color white. In this curriculum, it is associated with philosophy through rational/intuitive thought.

The West is the domain of the setting sun, the source of sustenance and social well-being and relationship within community. In this curriculum, it is associated with the domain of social psychology and self-knowledge. It is the dwelling place of the self and the group mind.

The South harbors plants, good fortune, spiritual richness and the fertility of the earth. It is the place of daylight, the full sun and the warm, feminine fruit bearing spirit. Its characteristic process and product is symbolized by the health of the whole. In this curriculum, it is connoted with the domain of medicine, and the quest for health and wholeness.

The North, the domain of animals, is the night, the unconscious and the unknown. In mythological terms, it is the place of the cold dark wind, origins and internal space. Its characteristic process and product is symbolized by the primal instinct, myth and dreaming. In this curriculum it is associated with animals, mythology, and the quest for understanding "the animal within" and the archetypal unconscious.

ETHNOSCIENCE FROM A NATIVE AMERICAN PERSPECTIVE:
A CURRICULUM MANDALA

Below (Nadir) is the domain of the earth mother and the archetypal elements of earth, fire, water, air and ether. This is the place of earth woman, mother of the winds, the waters, the earth mounds (mountains), the fire inside the earth, and the breath of life and thought. Its characteristic process and product is the dynamic interplay of the archetypal elements in geophysical processes such as weather, volcanic activity, erosion, plate tectonics and bio-regional ecology.

The "Above" (Zenith) is the place of the Celestial Father, the Great Mystery and the ultimate expansive spirit symbolized by explicate order in the cosmos and man's quest for an understanding of the Universe and the Universal.

This basic idea has ramifications that go very deep in the cosmology of each tribal group. A cosmology is a culture's guiding story, and that story reflects on ways of relating and understanding that culture in natural community. I began to evolve models and build conceptual orientations which I felt were true to the universals I had come to understand with regard to Indigenous people.

The first course in the model helps students in all other learning that they do. This course not only facilitates their learning in this science curriculum; it also facilitates their learning in all the arts, because the first step for every artist is to understand the basis of his or her own creative process. Indeed, the first step for any endeavor is to understand. Unfortunately, this is not the way mainstream curricula are usually established or evolved. Their purpose may be to impart knowledge, but not necessarily to develop the context through which that knowledge can take hold. As a curriculum developer, I am aware of the weaknesses of current approaches to curriculum development for Native American learners and it is this awareness that formed the impetus for creating a new perspective for guiding the curriculum development process.

After students complete the first course, they enter a preparation phase in which they prepare for a journeying process that is not too different from what indigenous people call initiation. They prepare through story, through experience, through encounters, for what they are searching is knowledge and understanding in a personal and direct way.

Native American cultures have traditionally oriented aspects of their lives with the four sacred and symbolic cardinal directions. The East is associated with the first rays of the sun on the new day, the metaphoric symbol of first insights and first understandings. I associate this orientation with a course that I call **Native Philosophy**. In that course we explore different kinds of indigenous philosophies related to how the world was created, how things came into being in the natural world, not only Native American philosophies, but also those of other indigenous peoples around the world, compared with Western philosophies. Students must understand that there are differences in orientation which relate to the philosophies people hold about the way they learn and the way that they define themselves. Philosophy is tied to something even more primal, the guiding story that is not only spiritual and religious but forms the frame of reference for a people and how they understand themselves. Ceremonies and rituals and other ways of a peoples' worldview are a direct result of that cosmology.

Exploring orientation, they begin to understand the beginnings of thoughts and ideas and concepts that are reflected in different things that people do. They reflect on why science is, why western science is the way that it is, the way that it works, and why it is formed and structured in that way.

Students see that indigenous cosmologies do not separate the human being from the rest of nature. In Christian cosmology God created the world, and after that he created man to dominate the Earth, thereby establishing a separation between the human being and the rest of nature. Students begin to understand that much of what they have been exposed to in mainstream science education describes a separation of the earth and its human denizens. I try to help them understand the cosmological as well as the philosophical bases for this within the Western tradition. Many begin to understand why they have had so much conflict through their schooling with the whole notion of Western science. They discover why they've felt alienated, and why they may have difficulty accommodating the underlying cosmology of Western science.

It is never a question of who's right or wrong or who has been more accurate in their description, but of understanding a cosmology and its interpretation.

Students are left to choose, either to integrate what they need from both worlds or to decide the direction they accept or explore from that point on. The choice is theirs, and through that first course in the process of creativity and creative thinking, they acquire tools to work that choice for themselves. They are ready for the journey.

The West is the orientation of community. We are communal beings and as such we cannot exist outside of community. No person is an island. I call this course "**Social Ecology**: A Native American Perspective" and it is about the ecology of relationship and people in community. Students find out how community can teach and how it conditions people. We explore social structures and social ecology and tie those to Indigenous cosmology rooted in nature. We explore differences between western social organization and Indigenous social organization, give examples and reflect on those things. We then examine some difficult issues—conflicts between social psychology and sense of community which lead to abuses, like alcoholism. We explore what happens to a community when you take away the sources of its identity, and why certain Indian communities suffer so grievously because part of their traditional way of being is obstructed or degraded or dismantled. We explore social psychology, community action and politics, and we come to understand our communal selves more deeply.

Each of these courses has a life of its own. No course will be the same each time it occurs, because each depends on what students bring with them and how they stimulate each other and the teacher. We listen to guest speakers, watch films, do research. Students become very interested in ideas based on their own experiences. They explore the nature of alienation and why certain events happened in their community. These courses spur a creative process, and with the tools of the first course, they develop a way of working through and understanding. Their study of art stimulates; ideas for art come from reflection, and they use their art form to represent their understanding and to express their anger and frustration, as well as their appreciation. This course is important because it seems to be able to restore a community or tribal memory.

Its main function is to help students re-engender, or remember to remember where they and their people have come from and how their communities are formed, and why customs and traditions exist. This is not merely a teaching of history. The student who is grounded in his or her center as well as in cosmology, then moves into relational consciousness of tribe and community.

It is a natural evolution, a natural progression, almost like a controlled "nuclear reaction", although controlled so that you can allow such to energize and empower people. Think of this in concentric rings: first the center, the next the philosophy course, the east, and next the West and its accompanying course, Social Ecology.

The next course is called "**Herbs, Health and Wholeness**," and its orientation is the South, the place of healing wind and a harmonious yet dynamic balance between self, community and environment. This course begins with an exploration of the uses of plants for food or medicine by Indian people. We pick plants, relate stories about the plants we pick and reflect on them. The understanding we are reaching for is that not only did Indian people depend on the natural environment for their food and medicine, but they did so in practical and direct ways. They used plants for healing, a practical science. We enter the more abstract realm of health and wholeness. We explore 'deep' ecology and reflect on the nature of the human being in relationship to community and the natural world and how that works in the dynamic of wholeness. We compare this indigenous process and model of health and wholeness with the Western medical model of modern and allopathic medicine, which utilizes different approaches to fix the body mechanism when it breaks down. The body, a machine with replaceable parts, needs chemical change to affect a particular disease.

Students compare the relative strengths and weaknesses of cultural models of medicine. They can actually trace a system back to its source, discovering that the threads of creative process and cosmology and community weave together. They see a progression of knowledge, again like concentric rings. When you throw a small stone into a still pool you see rings radiate. The same is true with ideas. You come to know the path an idea has taken. The student sees that the health and wellness orientation of indigenous people relates to a particular Indian cosmology and its holistic, spiritual approach to living.

Through each of these successive rings students enrich their knowledge and experience. Each of these rings reinforces the knowledge and understanding that has come before, but approaches from a different direction. When we look at medical technology, we look at different possibilities. Students begin to understand that things are integrated, that things are whole. They understand the nature of plants and their communities, as well as human ecology. They gain an understanding of the cultural dimension of healing and restorative process. A plant community's health depends on the health of the plants themselves. Students gain insight into the nature of interdependence and through such insight understand how the wholeness of a community is dependent on the wholeness of individuals in that community.

You establish this premise at the beginning—that the healing traditions of indigenous peoples were not focused on individuals only, but on the health of a family and the health of a community. When one person in the family was ill, this affected all the other people in the family. The Navajo healing process of the chantways applies not only to an individual but to the family as a whole. The healing ceremony is performed because of an individual but also for restoration of the harmony and the cultural sense of community. It is the responsibility of the community to help an individual become well and re-harmonize with the community and with forces that person may be out of harmony with, usually particular natural forces or particular situations, spiritual, psychological or physical.

The course associated with the orientation of North is called "**Animals in Native American Myth and Reality**." This course explores the mythology of human beings relating to the animal world. Most mythologies we have created are about animals as animals are the other living entities with which we humans have had the greatest amount of interaction. Native American mythologies possess a huge body of animal stories, as we came from a hunter/gatherer life. These are peoples' first stories of how they came to understand their relationships with other living inhabitants of the earth.

In this course we explore the origins of mythology and the origins of story, and how in the reflection of the metaphors and forms of stories about animals, humans understood themselves, mirrored their own situations and habits and behaviors such as greed and envy, self sacrifice and a whole variety of other human traits through the antics and interaction with animals. This course also deals with biology and animal behavior. It is a combination of what we know from a scientific standpoint of animal behavior, plus some wonderful story material that enriches —a creative integration between science and understanding. As people learn about behavior, they learn about their relationship to animals in a twofold way, and through that process gain understanding that they may not have taking a course in animal behavior or ecology.

Remember your early experiences with storytelling or listening to stories? Do you recall sitting as a first grader at the feet of a teacher? You may now be asking how older students accept this method of teaching?

Remember that they have been conditioned, because stories teach one to listen and analyze and or make analogies between what is happening and what they may know from a scientific explanation of animal behavior. Stories play with ways of explaining: they explain reality, and science in and of itself is a story.

If science as story is a specialized way of explaining natural reality, story contexts the whole process of science and allows you to look at it in some very different ways than might be ordinarily possible. What happens in the course "Animals in Native American Myth and Reality" is an exploration of animals as depicted in both art and science. This orientation allows students to create their own stories as well to incorporate, reflect and analogize aspects of animal behavior.

The next course is called "**Archetypal Elements**," and is oriented from the direction of the Below. In this we relate analogies of myth and creative process to the earth processes, earth, wind, fire and water, and how those elements become the foundation for not only life on earth but also for the expression of the Earth's very being. We explore the mythologies related to fire, the nature of water and its importance. We talk about natural bodies, like the mountain and the sea, and we talk about processes like earthquakes, volcanic activities, and about how indigenous people view these things. We examine how earth processes were portrayed in myth and what roles they had in the cosmology of a people, and through that, begin to understand the basics of geoscience like plate tectonics, weather patterns and the water cycle. These aspects of scientific knowledge are integrated into reflection on geophysical forces and their expression in Indigenous thought.

We finish with the last orientation, the "Above." This course is **Native American Astronomy,** which reflects on the nature of the cosmos and the nature of mythology as a vehicle to show how indigenous people attempted to resonate with the natural world and to build societies and organize themselves using the model of the cosmos as their primary inspiration. Students gain an understanding not only of the basics of astronomy and how people recognized and oriented themselves to the sun, the moon, and the stars but also used these to guide them as they went from one place to another. Just as the Polynesians used their knowledge of star positions to guide them from island to island, Plains Indians guided themselves across the sea of the Great Plains.

As students complete their exploration of these seven orientations, they feel the wonder and the awe of being alive in a natural place. They see how Indigenous people reflected a very sophisticated understanding of natural process and established their relationship in the natural world. They experience and think about nature in some very direct and creative ways.

Summary: Key Points for Processing the Indigenous Science Curriculum

1. The Indigenous Science Curriculum facilitates putting young people in touch with their cultural selves and their inner sense for learning. It gives them a foundation through which they may most fully express themselves in relationship to their family, community and the natural world. The activity of the "community of learning", which evolves through the process of the curriculum helps students become more whole as well as contributing members of their community.

2. The process of the curriculum facilitates students' realization of the Earth as the ultimate source of human, plant and animal life. Students are provided with experiences which develop their innate understanding of "biophilia" or that intuitive connection which humans feel for other living things.

3. Through the process of the curriculum students discover the beauty and complexity of nature through their experiences, encounters and learning episodes inherent in this approach to learning science.

4. The activities of the curriculum help to bring students back in touch with their cultural roots, the land, plants and animals.

5. Students learn how various Indigenous people practically made a "living" from the land which they came to call home.

6. Students learn practical skills of science and living in sustainable relationship to the natural world through experience and observation.

7. Students learn science through the facilitation of inner-experience and self-knowledge. This process begins with **bonding,** which then leads to **trust**, then following a **storyline** as it relates to an aspect of science in a contextual framework of **sharing and caring, looking inward** and **self-reliance.**

8. Students are given the opportunity to learn science in relationship to a cultural perspective and a world view, which mediates between themselves and science learning. This learning is accomplished through:

   a. Exploring cultural roots.
   b. Developing a historical perspective and empathy for the practices of Indigenous science.
   c. Reversing history by building upon the inherent strengths of Indigenous philosophy and environmental knowledge.
   d. Participating in cultural activities where appropriate.
   e. Sharing of traditions when and where appropriate.
   f. Learning and practicing Tribal arts.
   g. Learning and playing Indigenous games.
   h. Appropriately exploring the "ecology" of science learning.

9. Students are exposed to the principles of science through exploring and experiencing the world around us. These principles are "absorbed" when students:

   a. Learn how to apply "all their senses" to science learning.
   b. Learn how to "connect" with the natural world and how to "perceive" connections.
   c. Learn to "tune in" rather than "tune out" the rhythms of nature.
   d. Learn through the "multicultural" expressions of science.

10. The following are practical guidelines for creating and implementing an effective "Indigenous education learning environment". The Indigenous Science Curriculum:

    a. Applies the principle that we come to learn and we **can** learn science from many different pathways.
    b. Teaches for connecting to a "sense of place", a homeland.
    c. Facilitates learning to appreciate the land by living on it.
    d. Creates an extended family of learning by including community members, both young and old adults.
    e. Involves Elders and "special" community members wherever appropriate.
    f. Works from a cultural context to make meaningful connections of science to students' lives.
    g. Teaches through authentic learning experiences.

h. Creates a foundation for cross-cultural understanding.
i. Develops a flexible schedule for learning.
j. Emphasizes sharing and giving voice and vital expression to one's thoughts.
k. Facilitates personal experience and achievement.
l. Develops a foundation for health responsible living.
m. Gives students practice in applying their leadership skills.
n. Introduces Western Science, cultural and environmental studies through immersion, observation, appreciation and exploration with all the senses.

PART FIVE...

Chapter Fifteen......The Power of Myth and Story

As an integral part of the teaching/learning process, serious consideration of myth and story is rarely given in most modern American educational contexts. Yet children thrive on the mythical perspective, and there is evidence that the expression of childhood creativity is primarily facilitated by a mythological perspective. Myths mirror truths through a unique and creative play on untruths and imagination.

Within traditional Native American contexts, myth and storytelling are regarded as vehicles toward true understanding. They are a primal way of presenting realities and truths. Models of behavior, the significance of ritual, the basic realities of human existence and natural creative processes are presented in this form of coded communication. As the mythologist Ananda Coomaraswamy so aptly states:

"The myth is the penultimate truth, of which all experience is the temporal reflection. The mythological narrative is of timeless and placeless validity, true now, ever, and everywhere .... Myth embodies the nearest approach to the absolute that can be stated in words" (Coomaraswamy 1962: 33).

Storytelling and experience form the foundation for much traditional Native American learning and teaching. Stories give focus to and clarify those things which are deemed important. Experiencing through watching, listening, feeling and doing gives reality and meaning to important Native American cultural knowledge. Combining story with experience, Native Americans are able to achieve a highly effective approach to basic education.

Through the process of telling stories, skills in listening, thinking and imaging are creatively molded. Through experiencing, the skills of knowledge application, observation and experimentation are enhanced. Myth has been presented as a view of science. It is important now to look at myth from the perspective of storytelling and experiential learning.

Myths offer a great diversity of expression among different Native American groups. Myths that have survived the test of time are often those whose message is both immediate and timeless, eternal realities as true in the present as they were at the creation of the myth. Myths can express their meanings through a rich and creative use of language in an oral tradition, the art of a storyteller.

Because many Native American myths relate the learner to paradigms of proper relationship to plants, animals and all of nature, as well as to the consequences of a poor relationship to nature, they provide a place to begin a greatly humanized discussion of the general areas and underlying assumptions of modern science. As a vehicle for comparison and contrast between Native American and Western mindsets, and because they provide themes which are relevant to modern science, myths can play a vital role in the "hologizing" of science education.

Myths are themselves a holistic form of communication. They appeal not only to the intellect and imagination but also, through their enactment in song, dance, theater, oral recitation or art, to the entire human capacity for experience. Through myth, the Native American cultural relationship to the natural world is made to live in both mind and heart. In addition, myths provide a vehicle for explaining metaphysical realities and mindsets encountered in all cultural sciences that are extremely difficult to discuss or explain through any other means. In all cultural systems of science it is these metaphysical realities and mindsets that provide the foundation for the way that system views and comes to understand nature.

Myths perform four basic functions. The first is to kindle and represent a sense of awe combined with the realization of man's relatedness to the natural world and the universe. The second is to represent or relate a mythical history of creation, how things came to be, how a pattern of relationship or a perspective of the natural world was first established. The third function lies in the structuring and representation of symbolically-coded cultural knowledge. The fourth function revolves around the development of imagination and representational thinking as it involves living a myth through its reenactment and application of its precepts (Eliade 1963:18-19).

The importance of the mythological perspective is multi-dimensional with respect to this curriculum. For instance, at one level myth, through the oral traditions of Native Americans, provides a way of communicating about nature that has seldom been surpassed by other modes of communication. Myth provides a vehicle for the transmission of generations of "understandings" concerning the natural environment. Myth provides a way to explain and think about natural phenomena which goes beyond the mere physical description of the phenomena, **a way to describe nature that combines actual observable physical characteristics with affective, psychological and cultural perceptions**. Northwest Indian cultural myths relate how a mythological being first taught them how to fish, the nature of the first fishery, and the way the people must relate to fish in order that they might perpetuate themselves and the fish upon which they are so closely dependent. Inherent in all Native American myths concerning the natural environment is a philosophy and the ethics guiding Native American behavior toward nature. The understanding, respect and conservation of natural resources, the land and all of life, is reflected throughout Native American myth (Hughes 1984: 5).

In summary, myths, through their symbols, images, metaphor and play on imagination are powerful tools in the development of creative cognitive abilities and in the development of a holistic perspective of science. Myths are themselves reflections of the creative process, and through their use the creative possibilities for teaching science naturally expand. Learning how to explore, interpret and relate a myth at its various levels of meaning is itself a kind of knowing which is transferable to scientific thinking.

Myths communicate meaning at a number of levels simultaneously. Levels of meaning and interpretation allow them to be used creatively for learning activities which directly relate to the exploration of a particular concept in science. For example, a learning unit dealing with the "Gaia Hypothesis" in systems ecology which characterizes the earth as a living super organism, can be introduced through the exploration of Native American myths relating to the Earth Mother concept. These myths are universal among Native American cultures and include "Changing Woman" (Navajo), "Spider Woman" (Hopi), "Thinking Woman" (Keresan), "Sedna" (Inuit), and a host of other representations. Myths also provide a way to compare, contrast, or integrate two ways of perceiving natural realities, and in doing so stimulate real appreciation of the aesthetics of science. In every culture, science as a system of thought is influenced and guided by the myth-making process. Indeed, science in modern society has itself become a major generator and molder of myth as it has become a focus in modern life. The possibilities of using perspectives of myth to enhance the presentation of science concepts to Native American students are, therefore, unparalleled.

In the telling of stories, the content of myth and everyday reality are integrated within the content of the learner. Stories kept listeners aware of the interrelatedness of all things, the nature of plants and animals, the earth, history, and people's responsibilities to each other and the world around them. Storytelling, like myth, always presented a holistic perspective, for the ultimate purpose is to show the connection between things. Through the cultivation of hearing, understanding and insight were enhanced by the stimulation of the imaging capacity of the mind.

Stories told about creativity—about how things came to be; they explained the what, why, and how of important phenomena; they related the myth behind the ritual; they described the way of healing, health, and wholeness; they presented practical information about how things are done and why; they illustrated and illuminated the universal truths and characteristics of human life. In all these dimensions, stories were rooted in experience and provided an intimate reflection of that experience. They were a way of retracing important steps in life's way and of developing an affective perspective of themselves, their people, and their world.

Storytellers fulfilled a vital role in the continuity of not only the tribal culture, but of the mindset concerning people's relationship to the natural world. In this respect, the storyteller was the philosopher-teacher of tribal America. That the storyteller earned widespread distinction in Native American cultures is no accident. Traditional Native American storytellers were masters of the art of making stories real through a variety of rhetorical techniques, creative dramatization and the skillful use of metaphor. The use of artistic symbolization, song and dance were commonly employed by traditional Native American storytellers to add flavor and emphasis to their stories. In many respects the role played by the storyteller is the forerunner of the more formalized and eclectic role played by the modern teacher today. Whether teachers realize it or not, every time they teach they are echoing an aspect of the storyteller's art. Storytelling, whether about science, history, social science, language, literature or art, is an essential dimension of the teaching process. Teachers must continue to learn about and express their innate potential in this area.

Storytelling, and its direct relationship to experience, plays a vital role in the presentation of this curriculum. A large portion of the traditional Native American content which can be utilized in this curriculum is in story form. Its full effectiveness as a teaching technique can at times only be realized through the complete retelling of these stories. The reason for this is, because more often than not, Native American traditional knowledge and perception regarding nature was contained and transmitted in this form. In addition, the story by its nature is a high-context form of communication which requires an equally high-context mode of transmission to explore its multi-dimensional meanings. The story format allows for great flexibility in both its creative applications and parameters of interpretation.

All stories have multiple levels of meaning ranging from the very basic and straight forward to the complex and the metaphoric. Stories, especially those of the mythic variety, present philosophical, psychological and ecological truths simultaneously. Such stories provide opportunities to analyze, explore and develop new perspectives about Native American cultural knowledge of the natural world. It is in these stories that the paradigms of the science thought process are outlined or otherwise exemplified. Storytelling and experiential learning are highly effective teaching techniques which, when used together, combine explanation with exploration and action.

However, both must be learned well and applied creatively if they are to be effective. There is an art to both the telling of a story and the facilitation of an experience. The art takes practice, though the basic techniques are quite natural and easy to learn. Their application requires only sincere practice to become basically effective.

The six steps to developing the ability to tell stories and facilitate basic experiential learning activity simultaneously are as follows:
1) Orient students to the story you are about to tell or the experience you are about to guide them through. Show or explain how the story or experience fits into the lesson, and how students can benefit from participating. Establish the relevance of the story or experience to their lives and environment.
2) Clearly explain the focus of the story or experience in as many ways as deemed appropriate.
3) Tell the story or facilitate the experiences with sincerity and a

positive attitude.

4) Through imagination, allow yourself and assist your students in becoming actively involved in the story or experience.

5) Reinforce the story or activity by emphasizing those elements which require it in new ways or by integrating them with other areas of teaching and learning.

6) Carefully plan, evaluate, discuss and follow up what has been presented (Heidlebaugh,T. and L. Littlebird 1985).

Although these suggested actions are easier said than done, they are the essential ingredients in developing a positive and receptive mindset to telling stories and experiences, essential in these teaching/learning strategies. A clue to knowing where one might begin is to examine the ways in which Native Americans traditionally told stories.

In Native American storytelling the relevancy of ideas is usually established by using names, places and concepts with which the listeners are known to be already familiar. The human voice is used in a variety of ways to dramatize, create the mood or otherwise bring to life the story. The delivery and the language used in the story are manipulated to fit the characters and scenes of the story. Verbal image making and metaphor are used strategically to help the listener image the story. A variety of methods such as prayer, song, or artistic visual aids are used to set the stage for the telling of the story, and to intimately involve the listener in what is occurring in the story. (Ibid)

Honoring the Essence of Story Learning:

By themselves events rarely have anything in them that we can understand. The wind blows, the tree falls, the ants eat the leaves: So what? It is through careful listening that we begin to understand what our place in all this is. What we learn we pass on, calling this "meaning." Storytelling is one way to organize meaning. It is how we put the world together. The events and their reactions become organized into a story, which is then communicated in the telling. The heart of this telling is a reciprocal activity, an equal exchange between the teller and the listener. Story Teaching is also the efficient organizing and communicating of meaning. (Ibid)

How do people learn from story? We know how we learn from teaching. People develop specific areas to be learned in mathematics or science or social studies and the material is put in an organized context that can be passed on. With story, we do something different. We do not focus on reaching an objective. Instead, we develop a context in which the imagination finds content, so what is learned is part of the whole.

We gain experience with story rhythm (expectation and resolution), balance (comparison, choosing, organizing, analogy), coordination (reinforcement through affective feeling and modes used together), connection (metaphor, relationship, coherence, objectives) and finally meaning (the holistic resonance with something deep within us).

An example of an impactful story is:

The Journey of Scar Face

The Blackfoot legend of Scar Face presents an archetypal hero's journey of spirit. The story of Scar Face is a teaching story that reflects the courage of an individual in overcoming obstacles of cosmic proportions. It illustrates the nature of the way Indigenous people viewed relationships with all things, people, animals, the earth, and the sky. The story is about "face," that is, the spiritual nature of character and learning how to develop our true selves. The story is also about journeying to the center, to "that place that Indian people talk about." This is a place of spirit within ourselves and in the world as a whole. It is in "that place" that knowledge and gifts of spirit can be obtained. It is a place of vision where one must learn how to seek. Its inherent message is found in the landscape of our souls and our wondrous universe.

With deep respect and honor for the Blackfoot way of "Indigenous" being, the following version of the story of Scar Face is presented. Scar Face lived with his grandmother because his mother and father had died shortly after his birth. His face had a birthmark that set him aside from all others and became the source of ridicule and shame. Because he was different, he was taunted by the children and whispered about by others in the tribe. As Scar Face grew older he withdrew and spent much of his time alone in the forest befriending and learning the ways of the animals he encountered. It is said that he learned to speak with them. And through them he learned how to be related with all things.

As Scar Face grew older he experienced all the things of life with humility and great reverence. He even fell in love, as young boys do, when they come of that age and express that facet of their face. The focus of Scar Face's affection was a young woman, Singing Rain, the chief's daughter. Singing Rain was also a special person, kind and with a gift of insight. Although all the other young men competed for her affection, it was Scar Face who she came to respect and love because of his honesty and good heart. However, when Scar Face asked for her to marry, she revealed her sacred vow to the Sun never to marry. This was her pledge of spiritual piety in the way of the Blackfoot. The only way she could marry was if the Sun were to release her from her pledge. On hearing this, Scar Face determined to undertake a journey to the place where the Sun dwells to ask the Sun to release Singing Rain from her pledge. And so, it is said that Scar Face began his visionary journey to the land of the Star People.

Scar Face did not know where the Star People lived, only that they must live in the direction the Sun set every evening, beyond the Great Water in the West. So Scar Face prepared himself with help from his grandmother, and when he was ready he set forth on his journey, a journey to the land of spirit. He first traveled familiar territory, but then began to enter into lands that neither he nor other members of his tribe had ever seen.

As the snow of Winter began to fall a hundred paths became open to him, and he became confused; he did not know which way to go. He met a wolf on one path, and with great humility asked for help and direction. Knowing the goodness of his heart, the wolf spoke to him and guided him to the right path. He traveled that path for a great distance until he came to another series of paths, and again he became confused. He stopped, set his camp and prayed. Soon a mother Bear and her cubs appeared on the path in front of him. Again, with great humility he asked for guidance from the mother Bear. The Bear spoke with great kindness and pointed out to him the right path. Scar Face followed the bear's path for many days until the path ended. Now there were no longer any paths in front of him to follow, only the vast expanse of the great forest. As he stood and pondered in front of the forest, a wolverine approached him. He called out, "good wolverine, my friend, I need your help." Again, he asked for direction and help from this friend. Knowing his heart and the nobility of his quest, wolverine responded with great kindness and guided him through the forest to the edge of the Great Water, where, exhausted, he made camp. He thanked the wolverine, and he thanked each of the animals that had helped him by offering them a gift of song and tobacco. He could see a twinkling of lights across the Great Water, and he knew that was the land of the Star People.

Scar Face did not know how to cross the water to "that place that his people talked about." But he was determined to find a way. Then two snow geese swam by and offered to take him across the Great Water. When they arrived on the other side, he thanked the geese and their relative for their kindness and great service to him. He made camp and then fasted and prayed for three days and nights. On the fourth day, a path of sunlight began to form in front of him leading toward "that place". He leaped onto the path and followed it as it took him higher and higher into the sky. When he reached the end of this path of sunlight, he came to a beautiful forest and another path, a path of great width as if made by thousands of people traveling on it for a long, time. As he followed the path he came upon a richly decorated quiver of arrows leaning against a tree. He wondered who they must belong to, so he waited to see. Soon, on the path coming from the other direction was an extraordinary looking Warrior dressed in richly decorated white buckskin. As the Warrior approached, Scar Face could see that this man was an image of perfection. He asked Scar Face if he had seen a quiver of arrows. In response, Scar Face showed him where the arrows were. Grateful and curious, the stranger introduced himself, "I am Morning Star." Then he asked Scar Face his name and where he was going. "I am called Scar Face, and I seek the lodge of the Sun." "Then come with me, Sun is my father and I live with my mother Moon in his lodge."

When Scar Face arrived at the Lodge of the Sun, he saw that the walls were painted with the history of all people of the world. Morning Star introduced Scar Face to his mother the Moon. As his father the Sun entered the lodge, a great light entered with him. Morning Star introduced Scar Face to his father Sun, the greatest chief. Scar Face was so impressed that he could not bring himself to reveal his reasons for coming to the land of the Star People. Sun and Moon treated Scar Face with great hospitality and asked Scar Face to stay with them as long as he wished. Over the next few days, Morning Star showed Scar Face the many paths in the beautiful land of the Star People. There was one path to a distant mountain that Sun had warned Morning Star and Scar Face never to go near. It was a mountain on the top of which lived a flock of seven giant birds that the Star People greatly feared.

One morning, Scar Face woke to find Morning Star gone. Scar Face arose and quietly left the Lodge of the Sun to take a walk and decide how he might ask Sun to release Singing Rain from her vow. He thought he might meet Morning Star and ask him for advice. As he walked, he began to feel that something was wrong, and the nearer he came to the mountain where the Giant Birds lived the greater his feeling became. He knew that there was some reason Morning Star had gone to the forbidden mountain.

Scar Face set out in search of Morning Star. As he climbed to the top of the mountain of the Giant Birds, he found Morning Star engaged in a ferocious battle with the birds. These birds were indeed savage and extremely large. They were about to Overcome Morning Star when Scar Face joined the battle. Scar Face fought valiantly and soon turned the tide of battle. One by one, Scar Face and Morning Star began to kill the Giant Birds until all seven were slain and their tail feathers taken by the two warriors.

Tired, yet proud of their accomplishment, Scar Face and Morning Star descended the mountain and returned to the Sun Lodge to inform Sun and Moon of the defeat of the Star People's most feared enemies. Sun and Moon were very impressed by the courage shown by both young men and were especially grateful to Scar Face for saving the life of Morning Star. In honor of the courage of Scar Face, Sun offered to fulfill any desire he would request. Yet, Scar Face could not speak his greatest desire. He remained silent until Moon, knowing his heart, spoke of Scar Face's love for Singing Rain and her vow to the Sun that prevented them from being together. Sun immediately responded by saying to Scar Face that he would release her from her vow. Sun touched the cheek of Scar Face, and the scar he had borne all his life disappeared. Morning Star in turn gave him special personal gifts and revealed to him that he was his spirit father, confirming the feeling that Scar Face had all along. Then Sun and Moon began to sing songs in praise of Scar Face and Morning Star. Sun and Moon then gave Scar Face many gifts, rich clothes, and a special shirt. In addition, Sun renamed Scar Face "Mistaken Morning Star" because now without the scar on his face he looked like Morning Star. Sun taught Mistaken Morning Star his own special dance, the Sun Dance. He said that if Earth People wished to honor him and bring health and well-being to their tribe, they should dance the Sun Dance each year when he had reached the highest place in the sky. Then Morning Star led his Earth son to the path called the Wolf's Trail (the Milky Way) and placed a wreath of juniper on his head. In an instant, Mistaken Morning Star was back on Earth and on a path leading to his own village.

Singing Rain was the first to meet Mistaken Morning Star as he approached the village. He told her that Sun had released her from her vow, and she knew in her heart from seeing and feeling the magnificence of him that they could now be together always. Mistaken Morning Star called the people together and taught them the rituals of the Sun Dance. He showed the women how to build the Sun Dance Lodge, and he taught the men how to conduct the sweat lodge ceremony and raise the Sun Dance pole. He taught them about the sanctity of their individual spirit and the nature of sacred visioning. He taught them from 'that place that the Indians talk about." [1]

There are profound lessons to be learned from stories like Scar Face. The traditional versions of the tale told in the Native language have a richness and depth of meaning that are difficult to translate. Such richness and depth are true of similar tales among Indigenous people around the world. They are like the mythical spirit deer: they leave tracks beckoning us, if we would but follow.

PART FIVE...
Footnotes

Chapter 13

1. Information pertaining to the actual physical process of tracking was obtained through conversations with Mr. John Stokes, a professional tracker from Corrales, New Mexico. However, the association of the actual process of tracking with that of the scientific process of investigation and the process strategy of the curriculum model is a synthesis developed by this writer through the means of analogy.

2. See Myerhoff, 1974, in the References Cited section for further information concerning the pilgrimage for peyote by the Huichol Indians. The Huichol pilgrimage, whose purpose is "to find our life," is representative of a metaphysical tracking process through a mythological and geographic landscape. There are many other examples of individual and group pilgrimages to "sources of life" among various Native American groups. As such, they provide an important area for developing insight into the holistic integration of myth, natural reality and process thinking inherent in traditional Native American mindsets concerning nature.

Chapter 15

1. The Story of Scar Face is paraphrased from the following primary sources: Grinnell, George Bird, "Scar Face". *Blackfoot Lodge Tales*, pp. 93-103, and Wood, Marion, "Scar Face and the Sun Dance", *Spirits, Heros and Hunters*, pp. 85-89

PART SIX...
The Learner Within Bicultural Education

Chapter Sixteen......*Teaching Native American Students*

A basic understanding of what constitutes natural reality and how best to establish communication about nature is the aim of bicultural science education.

Preliminary steps necessarily begin with a careful study of how students actually perceive familiar natural phenomena. One may find a mixture of observations based upon combinations of folk, experiential and school-derived sources. Such observations may appear contradictory and one might wonder how these disparate combinations of ideas can be comfortably contained in a single student's perceptions. To a non-Native American observer, this mixture of perspectives may seem paradoxical.

We are all capable of having more than one internally consistent mindset concerning natural reality. Western scientific schooling often makes it seem otherwise, and such conditioning eventually stifles creative learning.

Opportunities to learn about or to practice the skills necessary for Western science are not present within many Native American homes. However, many students from a traditional background have had relatively rich experiences gained through a variety of cultural and practical encounters with the natural environment. The sources of knowledge of nature and the explanations of natural phenomena are often at odds with what is learned within school science or proposed by Western scientific philosophy. Herein lies a very real conflict between two distinctly different mindsets: **the mutualistic/holistic-oriented mindsets of Native American cultures on the one hand, and the rationalistic/dualistic mindset of Western science which divides, analyzes, and objectifies on the other.**

Which is better? Which is right? In reality they can be complementary. To be sure, the native way of perception must be given a contemporized expression. The western way must be "opened" to be more receptive and respectful of the potentials of an expanded perception that takes into account contributions of the subconscious and even possibly, in time, incorporates elements of the metaphysical.

Science educators have generally adopted an either/or attitude. Most science educators viewing a non-Western explanation of natural phenomena have decided that if it doesn't fit the Western scientific framework, it is not scientific. Since the earliest days of missionary education through the years of B.I.A. boarding school education to the present, replacing "the primitive beliefs" of Native Americans with "the correct ones" has been an integral part of the hidden curriculum. Such a difference in perspective has caused much conflict in Native American students, families, communities and schools.

THE PHILOSOPHICAL ASPECTS OF CULTURAL DIFFERENCE

The following model is an adaptation of work done by Edwin J. Nichols, Ph.D., of the U.S. National Institute of Mental Health. This model was originally presented to an international mental health conference of the World Psychiatric Association, November 10, 1976, in Nigeria

ETHNIC GROUPS	AXIOLOGY	EPISTEMOLOGY	LOGIC	PROCESS
EUROPEAN EURO-AMERICAN	HUMAN-OBJECT The highest value lies in the Object or in the acquisition of the Object	COGNITIVE One knows through counting and measuring	DICHOTOMOUS Either/Or	TECHNOLOGY All sets are repeatable and reproducible
AFRICAN AFRO-AMERICAN	HUMAN-HUMAN The highest value lies in the interpersonal relationship between humans	AFFECTIVE One knows through symbolic imagery and rhythm	DIUNITAL The union of opposites	NTUOLOGY All sets are interrelated through human and spiritual networks
ASIAN ASIAN-AMERICAN	HUMAN-GROUP The highest value lies in the cohesiveness of the Group	CONATIVE One knows through striving toward the transcendence	NYAYA The objective world is conceived independent of thought and mind.	COSMOLOGY All sets are independently interrelated in the harmony of the Universe
INDIGENOUS OF THE AMERICAS	HUMAN-MULTIVERSE The highest value lies in the balance of relations between humans, other beings and spirits of past, present and future	AFFECTIVE-ACTIVE One knows through activity, symbolic imagery and rhythm	CONCATENATE All elements and beings of the Multiverse are linked together	PANTHEISM All sets are dependently interrelated in the harmony and balance of the Multiverse

What kinds of measures concerning science education can be implemented to decrease the confrontation of a student's cultural mindset with that of Western science? First, the student can be introduced to the basic skills of science; through the use of familiar things or events, one can build upon students' innate interests and curiosity. Using this approach students become involved with science as a process of observing, classifying and collecting information and making generalizations with reference to phenomena they know about. Second, once students learn to apply these basic skills, they can be presented with a comparison of the way in which science as a thought process is exemplified in both their particular culture and that of the larger society. An analysis of symbols as they relate to the explanation of natural phenomena in both Native American culture and that of Western science should be undertaken. <u>In no case should one perspective be presented in preference to the other.</u>

In every culture, the thought process inherent in science attempts to relate derived symbols of phenomena to each other to develop a pattern of thought. And while many Native American students may not be exposed to or have developed skills required for established patterns of Western science, they are exposed to the process of making sense out of the **natural phenomena** in their environment. That is, they have developed skill in relating important cultural symbols of phenomena within the framework of what is meaningful to them.

The model or symbolic map of concepts representing what is important in a particular culture's natural reality is vital in forming the way members of that culture apply the science process and develop their mindset. Much of the communication concerning natural phenomena is **highly contexted** in Native American cultures. That is, information and communication concerning natural phenomena are presented in the context most appropriate to their purpose or through the use of symbolic vehicles such as art, myth or ritual. The actual relationships between natural phenomena are observed and symbolically coded in a variety of forms based on experiential knowledge. In contrast, Western science is low contexted in terms of both its communication and its processing of information. That is, information concerning natural phenomena is often highly specific, parts-oriented and presented out of the context in which the phenomena naturally occurs or is observed.

What are other considerations for implementing a bicultural approach to science? First of all, that there is both an ideal and a reality in the implementation of any approach to education that directly affects the way in which one actually teaches science. If teaching is a communicative art, then one can apply the appropriate research concerning the teaching, accumulation and learning of language to explain the complexities of teaching.

Teaching is essentially the processing and communication of information to students in a form they can readily understand, combined with facilitation of their learning and relative cognitive development. Since language is the predominant mode of communication used in the teaching/learning process, studies concerning language acquisition become important for gaining an understanding of the dynamics of this overall process.

People appear to have two ways in which they learn a new language. The first is through a process of unconscious language acquisition, the second through a more conscious process. Acquisition, which characterizes the first process, is the most normal way of learning a language and requires no formal teaching. Rather it involves an immersion in the environment in which the particular language is spoken. The other method of language learning involves the formal process of studying the way a particular language is structured. This includes the learning of grammatical rules, correctness of form and other technicalities (Ovando and Collier 1985: 58-61).

Research has shown that when a first language is learned, that language is used as the basis for learning subsequent languages. We actively engage in a gradual, subconscious and creative process in which we acquire the knowledge and ability to use a language and its underlying assumptions and cultural frame of reference.

If one views science as a special kind of language for communicating information about nature, then the way a language is learned has important implications for the teaching and learning of science. Just as young children naturally acquire a whole language system by being in an environment in which that language is cultivated, science can be seen as being similarly acquired. This implies that for children to begin to learn science as a process of communication, they must be exposed to an environment which is acquisition-rich in reference to the "language of science." Ideally, both the home and school environments should offer many opportunities to practice and develop the application of the science process.

This is, however, rarely the case. The task then becomes one of creating rich science environments in schools, including various opportunities to encounter the natural environment. Some exemplary ways in which this can be facilitated are: field trips, visits to appropriate museums, national or state parks, art, social science, or culturally-related projects dealing with the science process, storytelling or guest speakers, hands-on activities involving science as process, and the creative presentation of science both as a discipline and a cultural system of thought (Ovando and Collier 1985).

The acquisition of literacy has been described in the "relevant input hypothesis" proposed by Krashen (1981). The input-hypothesis focuses on the idea that a key to acquiring the understanding of a second language is a source of content which is familiar, easy to understand, interesting and relevant to the learner. If science is a kind of literacy, the concept relates to science as a language with its unique content, symbol systems and structure which can be learned very much like a second language.

The input-hypothesis suggests that we acquire understanding through messages containing new structures rather than being taught them directly. The implications of this for the teaching of science are great. New perspectives of science can significantly alter the effectiveness of teaching science as a process and form of literacy. Students can learn "new messages" or the message of a particular aspect more effectively if it is based on their understanding of "new structures" related to what is to be learned. That is, one can teach about science by teaching about something else and relating that something else back, for example, by using content material from the arts, humanities or social science and integrating some of the ideas and structures contained within these areas into the presentation of science.

Physics is derived from the Greek word "physis", which means nature. Art is essentially "an applied study of physics", which attempts to express an aspect of reality as perceived through the eyes and psychology of the artist, using tools of art and a particular artistic medium.

On the surface art seems to be diametrically opposed to science. The scientist employs experiments, using quantifiable data in an attempt to understand and explain a physical relationship in nature. The artist uses metaphor, image and imagination to represent an aspect of internal or external nature or human reality. At the creative level, artistic creativity and scientific creativity are two sides of the same essential process. Both art and science are creative inquiries into the nature of world and cosmos. They both apply specialized symbols, tools, concepts and principles. And they evolve a specialized language based on a particular paradigm (pattern of thinking).

Given this inherent connection between science and art, it is possible to teach science using art and art to teach science. This foundational connection is applied in a curriculum process model entitled *"The Private Eye: Looking and Thinking By Analogy"* by Kerry Ruef. The Private Eye is a guide to developing the interdisciplinary mind, hands on thinking skills, creativity and scientific literacy. In this model art activities, such as drawing, composing, observing creative writing and application of artistic principles, are utilized to help students "to develop the main habits of mind of the scientist: looking closely at the world, thinking by analogy, changing scale and theorizing. To enter the process of the real scientist...to emerge a scientist...one who loves and can read the world" (Ruef, 1992: p139).

There are other aspects to be considered, factors which have definite ramifications for both the emotional and cognitive growth of the Native American student within the school environment. For practically all Native American students, school represents an emotional challenge. The variations of culture patterns and relative levels of acculturation which students bring with them from home coupled with individual personality differences combine to form an important component of their emotional structure. Maladjustment to a school environment has usually been blamed on the student's home environment. The values, religion, community and social context from which a Native American student derives a frame of reference are essential to understanding the way in which teaching/learning activities will affect the student. Even the styles of nonverbal communication utilized within the classroom and the social context of the school play important roles in molding a student's perceptions of education.

Cultural mismatch between home and school has been the subject of much research with direct application to Native American education. For instance, it has been found that often the way a particular group perceives itself as viewed by the dominant culture has an influence on the self-concept minority students have of themselves within a school environment [Ovando and Collier 1985].

Many Native Americans view themselves as not being part of mainstream culture because they are deemed as such by the school. That they are looked upon as being different has a detrimental effect on their self-image. As they grow older, they will begin to perceive what is valued and not valued within mainstream culture, and when they find that their core values are not valued, they will then either try to adapt or to retreat.

The perception of implied bias often has a direct effect on their attitudes toward certain school disciplines such as science. The complex interplay of this perception between student, home, school and community has often resulted in a manifestation of the "self-fulfilling prophecy" in the way some Native American students have adapted to their school environment as well as the way the school has adapted to them. **These students internalize the assumption based on their experiences in school that teachers, school and society expect less of them and as a result they expect less of themselves and adopt a stereotypical image of themselves and their cultures.** Therefore, great effort must be applied in encouraging and expecting excellence from Native American students.

Teaching is a communicative art and language its most basic element, so the way in which language is used in the presentation of content becomes very important. Native American students may come to school with a different orientation to sound/symbol relationships, and in some cases, different patterns of thought and styles of communication than that of other students and school teachers and administrators. These differences require a sensitive approach to the presentation of each subject, especially of modern science, as this is an area that may be the least familiar to Native American students coming from a traditional home.

In contemporary Native American cultures, where traditional culture and language are being revitalized, it is common to meet students as well as parents who are consciously involved in relearning or reviving aspects of their cultural heritage. Language revitalization, along with a resurgence of cultural identity, will have a direct effect on perceptions and attitudes students have toward science.

Even when a student does not come from a traditional background, the bicultural approach presents important advantages. Very much like the learning of a new language, the learning of science can provide valuable perspectives. The comparison of a particular Native American view of science with that of Western science can have the direct effect of broadening students' perspectives of science. It can help all students become more open and less isolated within the confines of one cultural view point.

Discovering and understanding the student, culturally, socially and individually, is a first step in implementing a bicultural approach to education. This study of student characteristics is often given only "lip service," and neglected or poorly represented in the development of curriculum. It is easy to alienate rather than motivate students if, in presenting a particular model, one neglects students' feelings about a particular approach. The record is replete with examples of researchers and educators modeling programs based on an "ideal image" which has been unconsciously colored by their own bias and backgrounds. The tendency is to follow models too literally and to overemphasize an ideal picture of a cultural group and thereby perpetuate stereotypes which do not exist in reality or do not reflect the evolving character of a particular cultural group.

Such stereotyping is often the result of relying too heavily on ethnographic descriptions while neglecting the fact that cultures change and that students may have a very different view than is commonly represented in the literature. While such descriptions provide an essential starting point, they should always be tested against reality. The best way to do this is by facilitating discussion of characteristics which students perceive as being a part of their culture as they experience it. Experienced reality may be a collage of values and perceptions which do not reflect the statements made in the literature. The student's reality does not negate traditional realities but rather exists beside or intertwined with these realities.

Getting reliable information on the cultural characteristics of students is essential to an effective and meaningful implementation of the bicultural education approach. Careful observation of student compositions, informal discussion with students and parents, and involvement with community cultural activities are all helpful in developing perspectives needed for bicultural education.

Learning style is a dimension of the ebb and flow of "inside" and "outside" realities conditioned by individual and cultural environments. One's learning style is characterized **by ways of thinking, ways of feeling and inherited tendencies**. Of these three, the affective, or ways of feeling, is the least well understood, yet at all stages of learning, it is one of the most influential [Pepper 1985].

Learning style comprises: "a combination of environmental, emotional, sociological, physical and psychological elements that permit individuals to receive, store, and use knowledge or abilities" (Dunn 1983: 5).

The home learning environments of many Native Americans are characterized by such factors as freedom of movement, learning through direct experience, hands on and activity oriented learning. These learning models emphasize visual, spatial and kinesthetic orientations. By contrast, in the typical school environment, free movement is significantly restricted and indirect intellectual learning — which emphasizes verbal, mathematical, and logical orientations — is the norm. The disparity between home and school environments is so great that some students experience a kind of culture shock which significantly affects their attitudes toward school (Pepper 1985).

While no one of the many possible behavioral learning styles has been isolated as being distinctly Native American, some general tendencies have become recognizable. These include: a predominant non-verbal orientation, a tendency toward visual, spatial and kinesthetic modes of learning, heavy reliance on visual perception and memory, a preference for movement and activity while learning, and a preference for process learning which moves from concrete examples to the abstraction (Pepper 1985: 21).

The recognition that a cultural difference of affective learning style exists between the home and school environment is an important step toward more creative and effective teaching strategies for Native American learners.

The following areas present other important considerations for knowing students and implementing a bicultural orientation to the educational process:
 • An exploration of students' homes and cultural background. This would include such areas as social orientation, parents' expectations of school, parents' educational background and affective orientations toward home and community.
 • Observation of students within the school context with a special emphasis on interactions with peer group, emotional characteristics, styles of verbal and non-verbal communication, and predisposition toward specific teaching or learning styles, such as predominant relational or analytical.
 • An exploration of student values which reflect cultural mindset. The goal of such an exploration would be to find those values which can be focused upon in the development of curricula which students perceive as being relevant to their cultural identity.

The dimensions presented here are preliminary indications of possibilities and considerations of the learner within bicultural education. Each of the areas have been addressed only in general terms. A comprehensive exploration of each area would be itself a major research endeavor which would not only enhance the understanding of bicultural education, but would also broaden the realm of possibilities for creative teaching.

The Teacher and Curricular Change

To affect a systematic process of curricular change toward bicultural education teachers must familiarize themselves with the research related to teaching Native American students. This familiarization should begin with a basic understanding of the history of American Indian Education and its current practice. The following observations by Cleary and Peacock (1998) personify some of the key understandings teachers of Native American students should know.

Attempts at education of American Indians by Euro-American governments and churches were characterized by overt assimilation through colonization and relocation under political control, suppression and replacement of language, religious conversion and assimilative economic reconstruction. Euro-American colonial authorities relocated American Indians into controlled communities well apart from Euro-American settlements. Indians were forced to learn the language of their colonizers and forget their own. Conversion to Christianity was a fundamental aim of all educational attempts. Indian forms of economics were forced to restructure to fit the agricultural, industrial and labor orientations of colonial society (Reyner 1992: 33-58).

Yet there is a traditional form of American Indian education characteristic of each tribe which shares principles, beliefs and orientations with all others. A striving for harmony seems to be central to most tribes as a foundation for teaching and learning. This characteristic is founded on the understanding that all things are related and that humans are interdependent with all other living and non-living aspects of the world (Cajete 1994).

The dissonance that many Native American students feel with regard to school experience is that there is rarely a direct connection to their predisposition for harmony and balance, as school focuses almost solely on secular understanding.

When an atmosphere of "negativity" exists at any level of the teaching-learning relationship, it registers with students and will affect their learning negatively.

American Indian students can have a different way of being compared to non-Indian students, as a result of subtle cultural constellations of values, perceptions, views and ways of interacting and expressing themselves as well as differences in language and patterns of discourse. This different way of being can be disconcerting to teachers who are unfamiliar with Indian ways of being. Examples of these "ways of being" are their tendency toward quietness, reserve, respect, doing things in their own time, not wanting to stand out in their group, not interfering with the activity of others, or watching and not trying something until they are sure they can do it (Cleary and Peacock 1998: 21-58).[1]

Traditionally, Indian youth are expected to be behave well, respect living things, land, community, elders and always show gratitude, generosity, courage, tolerance, patience and acceptance. Youth who do not conform to these cultural rules of conduct are eventually shamed, censored or shunned. Children were encouraged to watch an activity until they could reproduce the activity competently. This is why many Indian students do not like to be put on the spot by the teacher. If they know something they will demonstrate it through group interaction rather than individual reaction (Ibid).

Related to this is the privacy that Indian people may feel with regard to certain aspects of their language and culture. When selecting content to use in the delivery of science instruction, always select content in the light of knowledge and understanding of the students and the community in which the teaching is taking place (Ovando 1988: 223-240).[2]

Becoming knowledgeable of student learning styles is strategic to the full development of this curriculum model. The philosophy behind learning styles is predicated upon the premise that a person responds to educational practices with a personal pattern of actions and performance. Three categories, the cognitive, affective and physiological, have been recognized by researchers as being predominant among most learners (Irvine, J.J., and D.E. York, 1995: 484-497).

"Cognitive" refers to the way learners prefer to receive and process information and experience, how they create concepts and how they retain and retrieve information. An "affective" style refers to the way a learner applies interpersonal skills and persists in acquiring new information or processing experience. "Physiological" styles refer to preferences for learning and processing which stem from such factors as gender, physiological rhythms, nutritional and health factors. (Ibid)

The cognitive style is most often the focus of research in teaching and learning although the affective and physiological styles are now becoming a major focus as a result of the brain patterning research. Ultimately, learning styles are the result of the complex interplay of all three orientations mediated by context and cultural milieu.

Cognition is the mental process by which knowledge is acquired. Everyone acquires knowledge in their own way. Cultural orientation is a major influence in the way each of us comes to know the world. Indeed, culture is one of the most important filters for Native American students.

One of the earliest forms of cognitive style research is related to field-dependence and field-independence. These contrasting orientations refer to the way individuals respond to confusing or unfamiliar situations. Field-dependent or field-sensitive learners are individuals who favor working in groups to acquire new knowledge rather than individually. They conform to the predominant social cultural framework of their group. Field-independent learners are individuals who prefer to work independently in acquiring new knowledge. They are goal oriented and focus on learning for personal meaning. (Ibid 487-489)

Field-independent behaviors can be observed when a student shows that he likes to work alone, is task oriented and enjoys competition and obtaining recognition. In interaction with the teacher field-independent students seldom seek personal contact and are more formal in their communication about academic work. The field-independent student enjoys undertaking new tasks without the teacher's assistance, is eager to start new learning tasks and seeks non-social rewards.

Field-sensitive behaviors can be observed in "peer relationships" when the student likes to work with others, likes to help others and is sensitive to the feeling and opinions of others. In personal relationships with the teacher field-sensitive students exhibit positive feelings for the teacher and are interested in the teacher as a person. The field-sensitive student follows direction and demonstration, pursues rewards which fortify relationship to the teacher and is greatly motivated when working one on one with the teacher.

Native American students as a group are considered field-sensitive learners. However, there are many learning styles among them and these characteristics should always be used only as a general orientation:
   a. They prefer visual, spatial and perceptual information to verbal;
   b. They prefer to learn privately rather than in public;
   c. They prefer to use mental images to remember and understand words or concepts rather than word associations;
   d. They prefer to watch and then do rather than to work by trial and error;
   e. They have a well formed spatial ability;
   f. They have a generalist orientation and are interested in people and things;
   g. They prefer small group work; and
   h. They prefer holistic presentations and visual representations (Irvine and York 1995:490-491).

Generally, Native American students are intimately involved with their families, culture and community. Teachers must be aware of the community and cultural orientation from which their students come. **Native American cultures are essentially spiritual, oral, nature centered, tradition-based and communal**. These cultural orientations which may be incorporated into the classroom should be well researched. The following suggestions may serve as a beginning for such research:
   • Get to know the norms and values of the community from which the students come;
   • Be aware of the students' background knowledge and experiences;
   • Discuss the students' learning style with them and help them to understand why they do what they do in learning situations;
   • Be aware of any pacing of activities within a time framework that may be too rigid;
   • Be aware of how questions are asked, and think about communication styles;
   • Consider alternatives for those students who do not like to be singled out from the group;
   • Provide plenty of time for students to observe and practice before

performing;
- Be aware of personal space boundaries;
- Organize the classroom to meet the learning preference needs, and encourage both cooperation and independent activity;
- Provide feedback that is immediate, consistent, and private when necessary; give praise often and for specific achievements;
- Consider the whole process of learning when planning for student learning activity and the application of whole language and thematic approaches; and
- Be flexible and realize that educational goals or standards may be attained in a number of ways, so provide students with adequate choices for demonstrating their learning (Cleary and Peacock 1998: 201-246).

PART SIX...

Chapter Seventeen......Wholism and the General Systems Theory

Every curriculum is a process which begins with a written strategic plan geared toward a specific outcome. This curriculum is presented in its "ideal form," the theoretical dimension. Its administrative, formal and experiential dimensions can only evolve through implementation.

There are, however, some general, prerequisite factors which I believe are important if a curriculum is to germinate. Parents and students and the educators who play a role in the implementation must be familiar with, and open to, holistic, humanistic and creative forms of learning. The administrators must be able to structure the organization in such a way that it is highly integrated, flexible and has a cyclic flow of information. Finally, the educational establishment must view cultural differences and biculturalism as an asset.

As a socio-cultural system, science is an integrated whole intimately interrelated with human activities and a process that can be presented from the non-Western high contextual/cultural perspectives in valid and internally consistent ways.

By recognizing cultural identity and creativity, the holistic curriculum can form the basis for a more positive conception of self, culture and science.

From the Native American perspective, science, traditionally speaking, is an abstract, symbolic and metaphoric way of perceiving and understanding the world. From the Western cultural perspective, science is essentially practiced as a rational way to solve problems

Cultural consciousness need not be inhibitive to unfettered scientific inquiry. In reality cultural consciousness adds to the understanding of the contextual application of scientific knowledge and technology. Students learning science in the contest of a multi-cultural 21st Century World must be personally aware of the social and cultural implications of their work in science. To deny them the opportunity to develop cultural awareness will perpetuate, rather than remedy the current imbalance and narcissism of science education.

My intent in this work is to describe a general science curricula which is culturally relevant and to show that Western science can be presented to Native American learners in a positive and psychologically rewarding way. I have attempted to integrate general aspects of the Native American cultural mindset concerning nature with the concept of modern holistic theory. It is interesting, though not surprising, to note that traditional Native American education methods and concepts about nature are directly complementary to the most recent right/left brain and creative process teaching and learning methodologies.

This curriculum approaches science from the perspective of general systems, attempting to help students see parts and wholes and understand the interrelationships therein. Hologizing, learning to look at the whole while simultaneously looking at its parts, is an important tool in balancing the predominantly partialistic orientation of the Western scientific method.

The Western scientific method through its process of taking things apart, analysis, hypothesis formation and objective evaluation, has been successful in helping us to understand some components of nature. Yet this partialistic approach to real life problem solving has not only conditioned many people to see and deal only with parts of problems, but has limited those people's ability to deal with problems which require an understanding of whole systems. Examples of such limitations have become all too apparent in many segments of modern society such as economics, the environment, transportation, medicine and education, where specific solutions addressing only parts of the system have inadvertently created other problems elsewhere in the system (Land and Land 1982: 36).

George and Vaune Land's transformational theory provides one of the conceptual foundations for this curriculum because it allows for the integration of the parts of science education and facilitates the understanding of thinking from a systems and developmental standpoint. The curriculum views science, creativity and culture as developmental and experiential process, and attempts to guide, facilitate and enhance the evolution of the scientific thought process.

Science literacy and the strategy of this curriculum involve the understanding of the organization, parameters and synergistic interaction of thought as they pertain to science. In addition, the curriculum attempts to reflect a dynamic, evolving and creative system of science education constantly involved in the process of taking in information from the environment, breaking it down, putting it back together in new ways, evaluating the outcome of its processing, implementing the best adaptation and responding to the reactions which the adaptation causes in the environment. These steps occur also in the creative process.

Another aspect of the general systems approach which this curriculum reflects is the evolution of the growth and understanding of science as a holistic system of thought by the student. As he or she becomes involved with each succeeding phase of the curriculum process and content, the student cognitively expands and grows.

Using the analogy of language to describe the learning of science, the Lands' **formative** first stage would involve the learning of basic grammars, paradigms, principles and symbols of the system. This entails the initial patterning of thought concerning a perspective of science through discussion, exploration and experimentation. The second or **normative** stage would include completion of thinking concerning the perspective of science followed by practice with the concepts and patterns learned, and the adding of new pieces of information to enhance further understanding. In the third or **integrative** stage creative experimenting with ideas, symbols, observations and materials relating to what has been learned would be explored. The fourth or **transformative** stage would first involve the purposeful destructuring of what has been learned in order to restructure it at new and higher levels of understanding. (Land and Land 1982: 37-44).

Thinking that is based on the dynamics of a general system must continually evolve. The evolution of thought and symbol systems goes hand in hand with creative growth and development. Through this kind of curriculum, Native American students can begin to perceive their cultures as involved in the dynamic process of creative evolution and become more aware of their active role in that cultural evolution.

In a dynamic system of learning, one learns as much, if not more, from mistakes and failures as from successes. Of further value is the understanding that in systems thinking, "feed forward," or new information needed to adapt to constantly changing situations, is just as important as "feed back," or reactions from the environment to one's actions. Inherent in this approach is the realization by students that systems exist everywhere, that the only differences are those of level and degree, and that all systems are similar in their reflection of the creative process of growth and development. Such an understanding is transferable to practically every situation students will encounter in interaction with their environments.[1]

Because general systems theory embraces not only specific classifications, but also philosophy and metaphysics, it provides an ideal vehicle for a holistic perspective of science. Students can begin to see science not only for its ability to look at but also to feel the natural world. The call of some educators to begin "humanizing science before it dehumanizes us" is answered through this approach. **But teaching science from only one cultural perspective and in a partialistic manner continues to be the central dilemma of science education today.**

PART SIX...

Chapter Eighteen......Systems Theory As Mirrored In Native American Ritual And Myth

Resonance with nature is mirrored by Native American art and in the aesthetic structure and expression of Native American ritual and myth. In both, one finds parallels of organization which reflect the principles of modern Systems Theory. These parallels can be said to be inherent in Native American ethnosciences in that they study cultural understanding of the way natural systems work. Because all traditional Native American cultures developed a close relationship to their natural environments, their ritual, myth, even social organization reflected the characteristics they perceived in natural systems. Nature provided the ultimate model for living and human expression.

Fritjof Capra in <u>Turning Point (1982)</u> explores Systems Theory in relationship to process thinking, communication and evolution. From a modern systems viewpoint, the world is seen as an infinite number of systems which are integrated through various levels of interrelationship. The way whole systems are organized and the nature of the interrelationships are the primary focus of systems theories.

Natural systems share some basic characteristics:
1) Natural systems everywhere "transact" in mutualistic interrelationships at a huge variety of different levels;
2) They are synergistic, which means that the whole of a system is always greater then the sum of its parts;
3) They maintain their stability and continuity through a dynamic yet well-coordinated flow of information through the various levels within the system;
4) Natural systems are self-organizing, especially as they pertain to living things. All living systems are continually involved in renewing and recycling of parts and processing of information in order to maintain their stability and overall structures. They achieve this dynamic state through the ability to reach beyond themselves through the creative application of their various traits such as the ability to adapt, evolve and learn in response to an ever-changing environment; and

5) Living systems are characterized by a state of equilibrium commonly called "homeostasis." This is a dynamic state of transaction in which a system has a great amount of flexibility because it has many options available for interacting with its environment (Capra 1982).

These five general characteristics of natural systems can be found reflected in a variety of ways in Native American ritual, myth and social organization.

The Spider Woman myths found in many Native American cultures personify these dimensions of systems process in a metaphoric form. In Native American mythology one often finds the theme of Spider Woman and her all-encompassing web. Among many tribes Spider Woman represents a kind of creative intelligence whose web metaphorically represents the innate interrelationship between all things. Many tribes believe that when any part of the strands of this cosmic web are disturbed or destroyed, disharmony is felt throughout the web and eventually registers with Spider Woman.

Similar representations of this characteristic of systems can be found in myths pertaining to "Changing Woman" among the Navajo and "Thinking Woman" among the Keresan Pueblos. The point to be made is that systems thinking and process are an integral part of Native American conceptualization of the natural world.

Synergism, the flow of information and self-organizing characteristics of systems, can best be viewed in a Native American context by exploring traditional processes of social organization and ritual expression. The goal of all major Native American ceremonials is the reintegration and reaffirmation of community holism, a general attempt at re-harmonizing that Indian group's social order. This re-harmonization can occur at many different social levels starting with the individual, moving to the immediate family then to the extended family, then out to the clan and to the community as a whole. An observation of a major ceremony of this type offers many insights into how this phenomena and social drama unfolds, enfolds and transforms itself at all levels of social organization. Lewis Thomas in Lives of A Cell (1974) provides an analogy when he states that a single cell, a group of cells, an organ, an organ system, and the whole living organism, organize and interrelate at a number of unique yet highly integrated levels. All primal cultures of the world exhibit this similarity to natural systems

159

organization in differing degrees. It may indeed reflect an intuitive and very basic tendency in the primal subconscious of all (Thomas 1974: 11-16).

The profuse use of art, dance, music, song, prayer and meditation, as a way to communicate at the higher levels of being, dominate the acts of information transmitted in all Native American ritualistic activity. When language is used it is of a symbolic nature, which attests to an intuitive attempt to transcend or otherwise enhance the limitations of language. This is especially apparent at its highest level in the psycho-spiritual plane of experience, which is entirely non-verbal in nature. The universal use of poetic oral language as a means of communication at this level is also an attempt to enliven and add power to ordinary language in the form of prayer, song, oration, storytelling and mythology. Capra states:
"When one examines the transmission of thought in these forms of communication it becomes apparent that systems thinking is process thinking; form becomes associated with process; interrelation with interaction; and opposites are unified through the continuous and dynamic ebb and flow of all parts of the system." (Capra 1982: 268).

Self-renewal is an essential aspect of self-organizing systems. Capra states:
"A living system is continually engaged in breaking down and building up through never ending cycles. All these processes are regulated in such a way that the overall pattern of organization of the organism is preserved. This remarkable ability of self-maintenance persists under a variety of circumstances including changing environmental conditions and many kinds of interference" (Capra 1982: 271-272).

Native American tribes have shown marked resilience in the face of changes at all levels of their immediate environment and many kinds of interference. In this respect, Native American social organizations reflect a sort of communal homeostasis.

The process and relationships inherent in natural systems can also be represented by using the analogy of a systems tree. This analogy allows us to more easily visualize the dynamic operation of natural systems. The symbolism of the "Tree of Life" from the Native American perspective provides a useful metaphor. As Capra so eloquently states:
"as the real tree takes its nourishment through both its roots and its leaves, so the power in a systems tree flows in both directions, with neither

end dominating the other and all levels interacting in interdependent harmony to support the functioning of the whole" (Capra 1982: 282).

The tree of life is a mythological construct which occurs in mythologies worldwide. In Native America its significance as a focal point for healing and all ceremonies which revolve around "seeking life" is almost without parallel. It makes its appearance in such ceremonies as the Sun Dance and the Valadoras ceremonies of Mexico. Its meanings symbolize the cosmic inter-relatedness of all living things at their core essence. Living systems do resemble a tree, a cell, a human organism, and a human social community, because all of them are parts of the greater whole.

The organismic view of society is not new for it has appeared as a theme in one form or another in all branches of the social sciences. John Collier, in his early writings on social philosophy, regarded face-to-face groups as akin to organisms with all their aspects functionally related. Collier reflected the prevailing views of the social progressiveness of the turn of the century in that it was his belief that the real essence of society is cooperation and reciprocal obligation:

"Every element of the society — language, technology, value systems, symbol systems, world view — is functionally interdependent with all the elements of the society. Take one of these elements out from its functional place or role and you kill the element or kill its meaning and alter the meaning of the whole. Only within the wholeness of the society does any part live" (Collier 1963: 435).[1]

The feeling for the whole, "the People," is an integral part of the social psychology of all Native American and primal peoples of the world. This feeling for the whole permeates all social interaction within any Native American community. It is the basis for the traditional codes of ethics, political and socio-religious organization and activities.

PART SIX...

Chapter Nineteen......Teaching Strategies

Engendering creativity in the science learning process and the perspective of science as a cultural system of thought are two of the major foundations of this curriculum. The development of the curriculum requires the application of a number of important and interdependent teaching/learning approaches. These include:

1) An understanding and application of the metaphoric thought process;

2) The understanding and application of teaching and learning strategies which address both the brain patterned learning styles of students;

3) Teaching for creativity;

4) The development and application of situational learning contexts where there is a specific interface between science and culture;

5) The facilitation of opportunities for student growth and development in their abilities to deal with and adapt to changing environmental influences; and

6) An understanding and application of interdisciplinary perspectives concerning science, culture and creativity.

Each one of these areas involves a highly complex, process-oriented strategy which requires research and experience to administer effectively. Therefore, further research should be guided by the needs of the student, learning context and teacher in each specific situation.

1) *The Metaphoric Thought Process*

Metaphor is defined as "the application of a word or phrase (or symbol) to an object or concept that it does not literally denote, in order to suggest comparison with another object or concept" (Random House College Dictionary 1975). In this curriculum, the strategic use of metaphor is essential in that it provides a "connective bridge" between the imagination and fantasy faculty of the right brain on the one hand, and concept and reason faculty of the left brain on the other. Metaphoric thinking stimulates the development of cognitive abilities that are higher than simply reading, writing, or listening. The use and understanding of metaphor greatly enhances concept attainment, the development of insight and the ability to use imagery. Metaphors are highly flexible in their uses and can be adapted to virtually all cognitive levels, from child to adult. As such, the use of metaphor provides an important tool in the development of a culturally-based educational curriculum as it facilitates the use of cross-cultural examples to illustrate desired concepts.

The understanding and application of the metaphoric thought process is invaluable both as a teaching strategy and as a thinking skill which can enable students to dramatically increase their creative thinking abilities in the area of science.

The use of metaphor as a teaching tool is an ancient strategy that has been used by virtually all the great teachers of human history. That it is an integral part of storytelling and mythology reflects the great capacity of metaphor as a vehicle for conveying highly abstract concepts. Allegories, parables, riddles, visualizations, symbols, koans, poems, ritual and myth all provide specific expressions of metaphorical thinking.

"Many of history's great teachers have recognized the power of the metaphor and have intuitively perceived that the rational, verbal part of the mind requires more than just rules and guidelines. When we review the teachings of these past masters, we find they seem to know that directing the mind means capturing the creative insight of the mind — the part of us that grasps the implications of the concepts and remembers the relationship over the long term. Over and over these teachers used "right brain" images and metaphors to teach their followers social and political concepts (Sanders and Sanders 1984: 20)."

The intimate use of metaphor in mythological thought and perception is especially evident in Native American mythologies. In these mythologies, metaphor provides the key vehicle for the presentation and elaboration of cultural truths, relationships, behavior and personality traits deemed important in particular Native American contexts. This is especially the case in myths which relate concepts and ideal relationships to the forces of the natural environment and all the living things therein.

Metaphoric thinking is closely involved with the process of imaging in creativity. Metaphor allows for the expansion and elaboration of creative insight through the mobilization of right brain processes such as synthesis, intuition, symbolization and the process of relationships (Sanders and Sanders 1985: 5-7).

Its use provides a link between right and left brain processes and perception in relationship to particular principles. In providing this link, metaphor provides an invaluable aid to the internalization of a concept by a student. (Review diagrams for the Right and Left Brain on page 115)

2) Teaching To Brain Patterned Learning Styles

The enormous amount of research in the past three decades concerning brain hemispheric specialization has profound implications for modern education. Educators are only now realizing the potential such knowledge forecasts for future education. Much of the prior emphasis of modern education has been on developing only the inherent potentials of the left brain thought processes to the almost complete neglect of the potentials of the right brain. Educators must begin to address seriously the inherent potentials and characteristics of the "whole" brain.

This is not a new insight. Traditional Native American teaching/learning methods intuitively employed strategies which integrated both right and left brain processes in the act of learning. In fact, traditional education in cultures around the world have incorporated techniques which directly access whole brain thinking. The use of metaphor and storytelling by different world cultures is but one of the many examples of such whole brain techniques.

Characteristic duality of brain functioning can also be seen in world cultural art, architecture, mythology, religion, philosophies, medicine and science. The dual nature of human thinking has been illustrated in a variety of ways within such early approaches to education as the Socratic dialogue, the use of parables and koans in some forms of Oriental education, the Montessori method and Rudolph Steiner's education through myth. In Native American traditional contexts, the reflection of the mind's duality is reinforced through a variety of myth, philosophical teachings and even social organizations.

In all that we think and do the two halves of the brain work together. The right brain initiates the idea through its imaging, intuiting, and patterning capacities. The left brain develops the material through its capacities of analysis, abstraction and objectivity. The mutualizing/reciprocating interrelationship between the right and left brain processes is constant.

Whole brain orientation is prevalent in Native American cultural sciences and arts as exemplified in the almost complete integration of science with both art and religion. However this should not be interpreted to mean that Native Americans do not use left brain processes. To the contrary, Native Americans consistently reflect highly skillful use of observation, analysis, experimentation, categorizing, rational and abstract thought. The basic difference between Native American cultural sciences and modern Western sciences is one of culturally responsive emphasis in reference to which brain processes have played the predominant role in their respective evolution.

Traditional conditioning of Native Americans in reference to knowing and relating to the natural world continues to affect many Native American students.

The contrast of cultural mindsets reflects the differences between cognitive brain processes and provides a focal point for the strategies employed in this curriculum model. This entails not only presenting a different kind of content which creatively integrates Native American cultural and modern scientific perspectives but also involves how such content is presented in reference to the characteristic learning styles of Native American students. (See Diagram: The Creative Process Instructional Model on pg. 171)

The highly-structured drill and practice oriented approach to science education prevalent in American schools today must be balanced with an equalizing emphasis upon right brain teaching/learning processes. To this end, this curriculum model also incorporates teaching to right/left brain learning styles adapted from a model developed by Dr. Bernice McCarthy. The model is called "The 4-MAT System" and emphasizes teaching to learning styles with right/left brain mode techniques. McCarthy developed the 4-MAT system based on brain hemispheric specialization research. The model addresses a "felt need," which McCarthy observed as a direct result of her experiences as a teacher and her own tendency toward right brain processes.[1]

The 4-MAT model posits four types of learners, each characterized by a style of learning oriented to either right or left brain processes in varying ratios. Each learner responds best to a type of teaching based on the way each learner processes information.

The Type One learner learns best through personal involvement, learns by listening, shares ideas, perceives through concrete experience, processes information reflectively, and uses imagination and innovation. Learners of this type are characteristically involved in finding out the "why" of things; they are idea people and function through social integration and bringing unity to diversity.

The Type Two learner learns by sequential thinking, reliance on authority and through abstracting theories and concepts. Such learners are task-oriented, and are rooted in analysis, rational thought and ordered problem solving processes. These learners excel at creating and learning from concepts and models. Learners of this type are characteristically involved in finding out the "what" of things; they excel in traditional classrooms and much of modern scientific teaching is geared to their style of learning, as such students thrive where logical/rational thought and objectivity are prized. These students tend to strive for intellectual recognition in school and generally receive it.

Type Three students learn through practical application and active hands-on activity. They are primarily motivated by a need to know how something works, how it can be used, and the kind of products which result from a concept or model. They are most concerned with the doing and acting out of ideas within the context of real life situations. They involve themselves with strategies and are prone to building, constructing and other applied sciences. They are security oriented and tend toward practicing and personalizing what they have learned.

The Type Four learner is primarily an integrative learner, tending to combine and synthesize experiences and then applying what has been learned to new situations. This modified form of trial-and-error learning leading to self-discovery makes the Type Four learner potentially the most creative learner. He or she is always looking for new possibilities and is inherently very adaptable to changing environments. Type Four learners thrive on experimenting, acting and testing through experience. They are involved with the "what" and "if" of situations, and are at their best when they are carrying out plans or strategies which they have devised based on their innate sense of a problem.

Learning a new subject begins with addressing the question "WHY" which involves finding meaning, connecting and examining. Next learning involves answering the question "WHAT" which involves finding order, imaging and defining. This is followed by exploring the question "HOW" which involves experimenting, trying and extending. Finally, the question "WHAT IF" is posited which involves creating, refining and extending.

The recognition that there is more than one type of learner and more than one way to learn has been voiced in modern education for some time. Actual development of teaching strategies have been slow to develop. There are many reasons for this, the most predominant one being the fact that teachers graduating from teacher colleges have been conditioned by the system to teach to only one type of learning style. Although all four types of learning styles are commonly exhibited by students in every typical learning situation, traditional teaching strategies have ignored the needs and potentials of all but Type Two learners.

In reality, it is difficult to address the learning style needs of all students simultaneously within a specific presentation in a subject area such as science. It is always much easier to homogenize a presentation and have the student adapt to the lesson. This is possible to a certain extent since everyone exhibits some of the characteristics of each type of learning style. However, research has shown that while all students possess characteristics of each of the four learning styles, each individual will tend toward a certain type of learning as the result of personal, environmental and cultural conditioning.

Native American learners can be found in any one of the four categories of learning styles. However, they tend to fall predominantly in the domain of the Type One learner and secondarily within the domain of the Type Three learner. Most science education, from which Native Americans feel disenfranchised, has traditionally addressed the learning characteristics of the Type Two learner. With the new interest in creative learning, there is now some emphasis in science education on addressing the learning orientations of the Type One learner. While this new trend is important in establishing a badly needed balance, the learning needs of the other two styles must also be addressed. The content and teaching strategies of this curriculum, therefore, tend to favor the Type One and Type Three learning styles. Type Two and Type Four learning styles, while not neglected, have been given less emphasis.

Assessment of the learning styles of the students is a prerequisite toward maximizing the effectiveness of this approach. Various evaluation instruments have been developed. The instruments for learning style assessment developed by Paul Torrance (1980) and the L.S.I. (Learning Style Inventory) developed by David Kolb (1976) have been used most by the author to assess the basic learning style emphasis of Native American students.

Once the teacher has established a basic learning style profile for each student, the application of the appropriate strategy of teaching/learning can be based on the characteristics of each learner, the learning situation and the content to be presented.

Bernice McCarthy's 4-MAT system provides an easy method for the presentation of specific content and teaching activities that address Native American student needs in four different ways based on the characteristics of learning styles. McCarthy's strategy begins with the discussion method in which there is a great amount of teacher-student interaction. The teacher's emphasis in this first stage is on motivation and observation of student learning. Student experiencing of this stage revolves primarily around attempts to integrate learning experiences with sense of self. Learning here involves searching and finding activities. It begins with creating an "experience," using primarily right brain processes, and moves to "reflecting and analyzing an experience," predominantly a left brain process. In the second stage of the strategy, the teaching emphasis is upon information processing methodologies. The teacher is actively involved in "teaching," guiding and the organized presentation of information to be internalized. This is the traditional lecture, demonstration and discussion which has come to be synonymous with modern teaching. Students are involved with the formation of concepts and basic problem solving. This process begins with "integrating reflective analysis into concepts" (a predominantly right brain mode) and then moves to further development (elaboration) of the concepts of skills.

The third stage of the strategy involves the extensive use of coaching. The teacher becomes a facilitator of learning spawned by the information giving process of stage two. Students are primarily involved with the "practice and personalization" of that which has been presented.

In the fourth stage, teaching is characterized by facilitating "self-discovery" among students from what they learned through the processes of the other stages. Teacher and students interact in evaluation and remediation activities based on the needs and progress of each individual student. For the student, this stage is characterized by an integration of applications of what has been learned aimed at some sort of creative synthesis. Students engage in an analysis (left mode) of what has been learned in search of meaning and relevance to their lives. This is followed by an adaptation of what has been learned to new and more complex situations or experiences.

Human learning is a complex holistic phenomena wherein categories meld together and several different learning processes occur simultaneously at different levels. The 4-MAT model presents a beginning at teaching to whole minds.

3) Using the Stages of Creative Process as an Instructional Model

Teaching/learning activities should always be situationally oriented. The situation dictates the kind of teaching or learning style that is to be applied. The ideal 4-MAT model moves sequentially through each stage of teaching/learning activity, with balanced emphasis on right/left brain processes and completes the round in the presentation of each learning unit. In reality, learning situations and students reactions to learning are highly variable and inherently creative. It is therefore important to improvise when necessary on the implementation of any model according to the requirements of each situation and its creative possibilities.

One way to do this is to simplify and apply the four phases of creative process as an instructional template. This can be done by following a simple formula in the presentation of each lesson. Give the reason for the lesson and teach it using the appropriate methodology; allow students to react to or otherwise interact with what you have presented; finally let students teach themselves or others something similar to what they have learned. This simplified sequence allows students to perceive process, act and intuit based on what they are learning and practice discerning patterns in order to develop hunches.

A more comprehensive application of the creative process as template for instruction might involve the following.
In the stage of **First Insight**, students are involved with encounters, myth, metaphor, cultural content and experiences which help them form perceptions paradigms of thinking and a sense for the essential compo-

nents of what they are about to learn. In the second stage of **Preparation/Immersion**, students compare and contrast Native American and Western scientific perspectives facilitated through presentation by the teacher and their own personal research. Students are introduced to the principles, forms or classifications inherent in the science content they are studying. In stage three **Creating/Inventing**, students process creatively what they have learned and experienced kinesthetically by doing art, role-playing or through hands on experimentation. Finally, in the stage of **Evaluation**, students reflect on the symbolic nature of their learning by presenting and discussing what they have learned. Both students and teacher have the opportunity to evaluate the progress in learning and the work which has been accomplished.

In addition, through the vehicle of metaphor, one can develop an <u>affective</u> connection with the learning process. For instance one can develop the significance or meaning of what is being learned. An affective connection to the knowledge and wisdom of what is being learned can be established. An affective connection to the integrity of what is being learned can be illustrated. In the final stage, an affective connection can be forged that relates to the delight and the courage to explore which is inherent in self-discovery.

Process is the key in this approach to teaching/learning. And whether the process begins with the left-brain orientation (beginning with the theory and moving toward a worldview), or with the right-brain orientation (beginning with the view and moving toward the theory), the aim is to make meaning — to help students develop a resonance with what and how they are learning.

Overlaid upon this ideal teaching/learning process is the dimension of the curriculum which adapts various traditional Native American teaching/learning methods to contemporary science educational needs. Traditional Native American approaches to teaching/learning tend to emphasize right-brain modes of information processing, because Native American cultural sciences include metaphoric and symbolic approaches to perceiving and understanding the natural world.

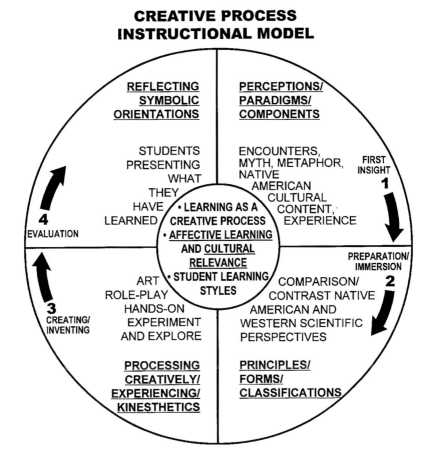

4) Teaching for Creativity

In his model of evolutionary creativity, George Land describes four levels of creativity, which relate to four stages of growth. The first level is "expansion" or simply the enlargement of an idea; the second, "mutation" or making changes toward improving the form of an idea while maintaining its basic function; the third is "hybridization" or the combining of ideas at a higher level to obtain a greater effect, and the fourth level is "transformation" or the act of invention or recombination of generated ideas into a higher level or relationship.

LANDS' TRANSFORMATIONAL CREATIVE LEARNING MODEL

PHASE	PHASE	DEVELOPMENTAL PROCESS	HUMAN LEARNING	CREATIVE PROCESS
PHASE IV	TRANSFORMATIVE	Search for New Patterns MOVING TO A NEW CYCLE	Learning and self become identified as a part of a larger reality meta-learning/consciousness	**Transformation** Disintegrating, surrendering, accepting, opening, building new perceived order
PHASE III	INTEGRATIVE	Sharing, destructing and restructuring COMBINING PREVIOUSLY EXCLUDED MATERIAL (exchange differences)	Synthesis or creating by combining differentness (general understanding thinking & relating skills)	**Hybridization** Abstracting, synthesizing, combining, metaphorical thinking and intuiting
PHASE II	NORMATIVE	EXTENDING AND IMPROVING THE PATTERN (likeness)	Specific knowledge and specialized skills. Practice through re-search, application and improving	**Mutation** Categorizing, comparing, analyzing, evaluating
PHASE I	FORMATIVE	ESTABLISHING A PATTERN (complementarity)	Basic survival information and skills. GRASPING THE "BASICS"	**Expansion** Perceiving, exploring, spontaneous acting

Diagram adapted from ("Forward to Basics" with George & Vaune Ainsworth-Land) Creative Education Foundation Inc.) 1982. D.O.K. Publishers, Inc. Buffalo, NY

The thinking involved in the first stage is simply that of perceiving and making appropriate connections in the solving of a problem. The second stage involves divergent and convergent thinking, generating lots of possible solutions, and zeroing in on those which are the most appropriate given the particulars of circumstances. Third level thinking involves the use of analogy, metaphor and innovation to solve a particularly perplexing problem. Level four thinking is transformational and involves the destructuring of thought and its restructuring at a higher level of understanding, a true metamorphosis of thinking in relationship to a problem (Land and Land 1982: 44-45).[2]

This model allows one to track the process of creative development as students are introduced to each level of creative thinking as it applies to science, they begin to see that for every problem they encounter, there is an appropriate level of creative thinking that can be applied.

The model reflects the intuitive understanding of growth and development that all humans seem to possess. Learners discover that it is important to come to terms with each phase of learning and creative thinking before progressing to the next phase. Experiencing the curriculum as a student moves from one stage to the next entails the experiencing of a creative process. In a real sense the student learns the nature of each of the phases of creating through "osmosis" without the process having to be formally explained by the teacher. Students learn about the creative process by being directly involved with a creative process. Once students begin to see different levels of problems and learn how to apply the appropriate level of creative thinking in search for solutions, they will have internalized one of the most important skills that science education has to offer.

The intent of teaching for creativity is to facilitate receptivity to creative thinking. Implementing the curriculum strategy requires that the teacher understand the requirements of each level of creativity and apply this understanding to the construction and presentation of each unit of study. A key consideration must be the differences in individual levels of cognitive growth and maturity that characterize each class of students. (Land and Land 1982: 60).

In the development and presentation of each unit of study conditions of creative learning must be considered.

In phase one, emphasis must be given to the "felt" needs of the student. Teaching is geared toward motivation, discovery, perception, activity, exploration and the search for "relevant" patterns.

In the second phase, analysis, verification, and evaluation followed by basic synthesis toward a more complete understanding of the content becomes the center of activity. The enhancement and elaboration of what has been learned or created is also a major activity in this level. Teaching activity at this level will entail the extensive use of examples supported by readings or other activities which augment and elaborate what has been presented.

Learning and creativity at its third phase involves a high degree of synthesis, metaphoric thinking, innovation, the widening of perspectives, the creation of new grammars and symbols and the destructuring and restructuring of content toward the goal of creating something completely new. At this level students come to intimately understand the internal dynamics of the content and tend to develop a greater sense of motivation and commitment to what they are learning. Teaching will involve the extensive use of metaphor and analogy in the presentation of content. At this level, teaching activities are aimed at motivating students in their own use of metaphoric thinking and creative synthesis. (Land and Land 1982:67-72).

This third level of understanding is the "ideal" goal of this curriculum, though it is rarely achieved in any structured learning environment. Each student is a composite of the dynamic internal and external characteristics which influence his living, learning, and creating and will always be at different levels of cognitive growth and development and each phase of creative learning will always have its own unique requirements. Teaching for creativity requires creative teaching at every level. Herein lies the ultimate challenge of implementing this curriculum.

In any class students will be at different levels of maturity and stages of cognitive and psychological development. Appropriate adjustments of the content and presentation of the curriculum to students relative to developmental needs is very much a part of this curriculum's processing. Some students need to be told everything while others wish to explore possibilities for themselves. As students grow and develop, they internalize their own relationships to what is being presented at each level of the curriculum. Through their increasing awareness of these relationships, they will become more aware of how different aspects within and without science are related to one another and thereby see the need for a process perspective in science. They also grow in their cognitive ability with the science process.

Students need to be encouraged to use differences of thought and perception while retaining the unique characteristics of each. Learning to combine differences resulting from cultural biases and modes of thinking increases the capacity of thought to include broader perspectives of nature and allows students to become aware of the nature of probabilities, becoming more comfortable with "uncertainties" within each system of thought. The realm of possibilities is expanded and science is no longer perceived as a closed system based on irrefutable facts that must be memorized. Rather it is rightly viewed as an open and potentially highly creative system.

5) *The Interface of Science and Culture*

Making science "real" to Native American students involves the conscious manipulation of situations and contexts to illustrate the relationship between Native American culture and the science concept being presented.

Modern science and technology have touched practically every corner of the earth and have affected every human cultural group. Native American cultures are no exception to this pervasive influence. The interface between culture and science is so pervasive that it is sometimes difficult to distinguish what is science and what is culture. Situations may arise in the workplace through the purchase of the products of science or by encountering science and technology through medical treatment. Within any of these situations, modern science and technology are presented without explanations and are used as a means to an end. These various encounters with modern science and technology can provide a wealth of content for the development of culturally-based science education. Science from this perspective is much more than simply a foundation for technology; it can form a basis for cultural exchange and understanding which is almost limitless because it is always expressed within a specific cultural context. Learning to recognize, understand and use these situations for the purpose of gaining a deeper understanding of science in process is a key consideration of this curriculum.

The expression of science and technology is situational. That is, science and technology are always applied as cognitive or physical tools in a specific situation for a specific purpose and in specific cultural contexts. The application of a specific cultural system of scientific thought is always relative. Therefore, the way one evaluates each expression of science must be relative to the context and situation. Each cultural system should always be evaluated from the perspective of that system and not on the basis of any other. While comparison/contrast is an important learning activity in bicultural science education, the notion that one system is inherently better than another must be avoided.

Through explorations students have the opportunity to focus upon the expression of the science thought process in a given aspect of a culture. For instance, with a visit to a Native American ruin followed by a study of the architecture of various Native American cultures and their relationship to cosmology, environmental requirements and engineering and then the same for modern architecture in American society, students can gain a greater breadth of understanding of the art and science within architecture. There are hundreds of situational activities possible which directly reflect the interface of science and culture. Different cultural reflections of science are presented from the perspective of those cultures and students are allowed to discover for themselves the differences and similarities between the expressions.

Situations can also be tailor-made to fit the requirements of a particular lesson. Although this is more difficult and requires more structure than depending on available examples of the science process, tailor-made situations can be the most useful in a structured learning environment. For example, if one is trying to describe the interrelationships between science, art and mythology among the Pueblos, one might begin by developing a slide collection of Mimbres and Anasazi pottery art. These ancient pottery designs incorporate highly stylized or specific representations of plants, animals and natural phenomena which reflect mythological and ecological perspectives. As basic cultural content they can be used as springboards to interesting and motivating topical areas. These might include a discussion of human ecology as exemplified by the ancient Anasazi and Mimbres cultures of the Southwest or the perceptions of natural phenomena such as the sun, mountains, water, wind, and rain. This kind of lesson development can be done for almost any Native American traditional art, mythology, or practical technologies. The key is creative synthesis.

The situational framework of teaching and learning also allows for a direct way of addressing the cultural identity of students. Where a distinct interface occurs between Native American cultural identity and an aspect of science, the student's sense of identification with his or her heritage is immediately mobilized. The student naturally becomes motivated to learn more about the nature of this interface and its principles. Peter Adler explains the importance of this identification,

"Cultural identity is the symbol of one's essential experience of oneself as it incorporates the worldview, value system, attitudes and beliefs of a group with whom such elements are shared…The center, or core, of cultural identity is an image of self and the culture intertwined with the individual's total conception of reality. This image, a patchwork of internalized roles, rules and norms, functions as a coordinating mechanism in personal and interpersonal situations."(Alder 1974:26).

What is learned and what is internalized in a subject area such as science is related to the way what is presented is perceived as being related to the core of a student's cultural identity. The motivation to learn anything is dependent upon one's personal and socio-cultural needs and characteristics.

6) *Student Growth and Development Through Adaptation and Change*

Every system of cultural science attempts to develop a basic understanding of change while searching for patterns in natural processes and developing classifications of natural entities. These classifications and recognitions are ever changing, evolving, or adapting as new information is gathered concerning the natural world. Science involves continuous exploration of the natural world moving through stages of growth and development. The knowledge gained from the science process evolves through each stage of development and continually transforms itself toward successively higher levels of understanding.

Numerous examples of growth and development of scientific knowledge in both Native American cultures and Western sciences are presented to familiarize students with these processes and to facilitate their awareness of their own growth and development in understanding their relationships to their culture and the natural environment.

Native American cultural sciences were characterized by adaptations to the surrounding natural environment and reflected a deep understanding of change, life cycles and creative process inherent in nature. Creative adaptation is exemplified in Native American farming, hunting, fishing, medicine, art, myth, architecture, material technology, religion, ecological practices, astronomy, and social organization. Through these examples students can be helped to a more comprehensive understanding of change than would be possible in standardized science curricula.

7) *Science, Culture and Creativity: An Interdisciplinary View*

One of the drawbacks cited by critics of science education is that it has a near-sighted view of itself and that little attempt is made to relate science to other subject areas. This tendency has significantly narrowed the possibilities for "cross fertilization" of ideas and perspectives with other areas of research. Students view explanations from the perspectives of other disciplines as less than valid. This attitudinal conditioning can present a major problem in the development of creative thought in science. The lack of an interdisciplinary perspective further alienates students from science and reinforces negative attitudes.

In reality, science interfaces with a number of other disciplines. It is the understanding and creative application of this interface that is essential in widening students' perspectives of science. An interdisciplinary approach presents science as it actually appears in the day-to-day life of students. Students become aware of the influence scientific thought processes and associated technology have on their immediate environment. This awareness extends to the realization that science affects practically every interaction that they have with their world.

Exploring science from the perspectives of such disciplines as art, psychology, history, philosophy and the humanities infuses it with new meanings and understandings. The interdisciplinary approach views science in relationship to the greater whole of human knowledge systems. This establishes the relevancy of science to the whole of human experience. For instance, exploring the historical evolution of the Western scientific paradigm and comparing this with the historical evolution of the scientific paradigm of a Native American group, such as the Aztec, provides a broader perspective and appreciation for both orientations than would be possible through any one discipline. Application of knowledge is gained from both disciplines bi-directionally, since learning flows in both directions.

Learning, like creativity, is not a linear, single-directional, process. Rather, it is cyclic, in a constant state of flux, multi-directional, and multi-dimensional. This essential characteristic of learning is more often than not disregarded in the way teaching and learning are traditionally structured within the academic presentations of science, art, the social sciences and humanities. This tendency of modern education lies at the heart of the charges of lack of relevancy and the ivory tower complex leveled by critics of contemporary mainstream education. The interdisciplinary approach forces a destructuring of traditional boundaries of disciplines and their restructuring and synthesis toward the goal of more complete learning.

Teaching, like learning, is cyclic, multi-dimensional and bi-directional. The teacher actually learns and grows significantly along with the students as he or she evolves through the curriculum process. The experiencing of the curriculum process and the interdisciplinary approach is in many ways just as important for the teacher as it is for the student. Through the dynamics of the interdisciplinary approach, the teacher gains fresh professional and disciplinary perspectives on what is being presented. The teacher is placed in a unique relationship with students as a co-learner, and acts as guide only in the learning of both the students and him or herself. This removes the sometimes stifling personification of the teacher as the authority and allows for a more intimate and dynamic relationship between teacher and student. Learning, relationship and understanding become the focus of interaction.

The combining of a cultural perspective such as that of various Native American cultures with current research in the areas of creative process, situational natures of learning and the nature of cognitive growth and development forms a creative and viable foundation for teaching science and other subjects as well. An interdisciplinary perspective taught with storytelling, metaphoric teaching, the mythological perspective and experiential learning is an approach which can work well for all areas of education. These methods have served Native Americans well for many generations. What is needed is a resynthesis of these methods into a form which fits the educational needs of Native Americans today.

PART SIX...
Footnotes

Chapter 16
1. Cleary, L. and T. Peacock, 1998. "Collected Wisdom." Boston: Allyn and Bacon Pub. Cleary and Peacock present a comprehensive study of best practices, thoughts and collective wisdom of sixty teachers of Native American student from throughout the U.S. This work is an excellent guide for teachers who wish to gain deeper understanding of Native American students and education.

2. From: Ovando, Carlos J. "Science," in Teaching American Indian Students,

Jon Reyner (Ed.), 1992. Univ. of Oklahoma Press, Norman, OK.

Chapter 17
1. Land, G.L. and V.A. Land, 1982. "Forward to Basics". Buffalo: D.O.K. Pub.

Land and Land provide a comprehensive overview of the developmental aspects of the creative process as it pertains to the transformative nature of learning.

Chapter 18
1. John Collier was guided in his administration of Indian affairs by a social

philosophy which reflected a view extensive in understanding of the holistic

nature of social groups. From this perspective, Collier tended to make decisions

only after weighing the holistic effects of such decisions. For further reference,

see Kunitz (1981).

Chapter 19
1. The use of McCarthy's 4-MAT Model in relationship to teaching to right/left brain learning styles is presented because it provides an excellent illustration of a teaching approach which relates to the thesis of this book. However, it is only one of many possibilities for approaching right/left brain learning. Another equally promising model is that of Susan Kovalik's (1995) Integrated Thematic Instruction (ITI) which is an instructional organizational model based on eight brain patterned principles of teaching and learning.

2. The use of the Lands' ideas concerning transformational growth and development in relationship to creative learning is presented because it provides an excellent illustration of an approach to teaching for creativity. However, the Lands' model of creative learning is only one of any possibilities for approaching the creative dimension of teaching and learning within the context of this curriculum model.

PART SEVEN...
Building Bridges of Understanding

Conflict between Western and Native American perceptions and understanding of nature arises only when one system of conceptualization is emphasized over the other. What I have attempted to show is that a bridge of understanding can be constructed even between two such strikingly different cultural mindsets as those inherent in Native American ethnosciences and modern Western science. Native American cultures reflect the science thought process systemically as shown by the integration of science/technology with art, religion, ritual and myth. Western science reflects the science thought process primarily through focusing on the individual parts of natural systems. The construction of a bridge between two such mindsets concerning the natural world affords the student of science a viewpoint and orientation which allows for a broader and more realistic perspective of science as a whole process. In doing so, it allows students a greater opportunity to develop an appreciation of science as a highly flexible and creative tool for understanding the natural world as well as their own relationship to that world.

General systems theory, creativity as a learning process and science as a cultural system of thought are lenses through which students can see how parts of natural and human social systems interrelate among themselves and within greater wholes. These bridges between two systems of science necessarily involve curricula which views both as valid, internally consistent and complementary systems of thought and action in relationship to nature.

The curriculum will have different meanings for different people, necessitating development of different grammars or ways of communicating these meanings. Students, then, are placed in a situation where they must develop their ability to respond to the different relationships that they perceive in Western and Native American cultural/scientific perspectives. Each student with the help of the teacher must develop his/her own strategy to make meaning out of the content encountered at each level of the curriculum. There is a constant ebb and flow, an unfolding and infolding, between all of the processes and components It is in these interactions and relationships that learning in science must be allowed to develop. (See Diagram for the Right Brain and Left Brain on page 115)

Human learning is a continually evolving and developmental process which culture guides. Culture presents the body of shared ideas, explains origins, presents ideals of behavior and defines social paradigms, the sociocultural design that human learning generally follows. Every cultural group creates its own design which it transfers to the children of the group via the processes of informal and formal learning situations.

Schools are a significant part of the cultural design which involves both formal and informal learning situations. Cultural learning within school may be in direct opposition to traditional cultural learning, resulting in conflicts between the home culture and that of the school with which students must learn to cope.

Student adaptations may take the form of alienation when the student withdraws from those elements or subjects which present the greatest discontinuity with the home culture. Some of the conflicts which result, however, can present opportunities for creative adaptation that allow students to become more aware of themselves and their potentials. In order for this to occur, educators must be aware of these discontinuities and must adjust both their approach and curricula to minimize the negative effects. Unfortunately this is rarely done.

Modern education continues to be a major source of discontinuity. Traditional learning and culture have been generally ignored which has oftentimes been translated by Native American students as a rejection of both themselves and their cultures. Students have reacted in a variety of ways which have included (to use the popular phrase) "turning off and tuning out."

The position of this writer is that science curricula must relate to Native American learners as relevant and meaningful. Science must be viewed as a cultural system, a way of communicating and a creative process if it is to fulfill its potential as an essential dimension of the cognitive development of students.

This curriculum model is tailored primarily to those schools and teachers which: serve or teach a number of Native American students, are experiencing problems in motivating Native American students in science and mathematics, have an expressed desire to improve the science literacy and preparation of their Native American students for science related fields, and are interested in integrating their regular science/mathematics curricula with aspects of social science, art and the humanities.

I might also dare to imagine that some of the concepts and precepts herein expressed could enhance the effectiveness of science teaching **in any school, to any student group**. Holistic thinking can help raise the level of Western science education as well as Indigenous science education.

Some of the potential benefits would include an opportunity to exercise creative teaching through the use of Native American cultural content and the highly flexible learning approaches presented by the model, a concrete mechanism for the integration of science/mathematics with aspects of other subjects, ways to teach students <u>how to learn</u> by helping them develop and understand their own creative learning processes, and an understanding and appreciation of Native American perspectives on the natural world and their contributions to scientific and medical knowledge.

Schools adopting these ideas might discover that students are increasingly motivated to consider science related careers, are increasingly scientifically literate as measured by selected standardized tests, that education becomes relevant both for students and communities, and that science and mathematics become integrated with other subjects through the interdisciplinary approach.

The overall intent of this model is to focus on the Native American learner who feels alienated from science as it is generally presented and exhibits a predominately right-brain learning style. In addition, this model attempts to address the needs of the learner who is creatively inclined and exhibits a visual, spatial or kinesthetic orientation.

Benefits to learners would be manifold. A student would be more likely to explore areas of interest in science because of relevancy of course content and presentation. He or she would likely experience an enhancement of self-concept and his cultural identity through an appreciation for Native American cultural expressions and contributions to basic scientific knowledge. Each student would develop a more holistic perspective of the interrelationship between science, culture, art and the humanities. Finally, an increased science literacy along with the development of understanding and application of the student's own creative process and the creative process in the arts and sciences would be a natural outcome of the experienced curriculum.

Finally, for Indigenous people, the revitalization of Indigenous knowledge through truly self-determined education provides the most direct route for Indigenous sovereignty. Becoming open to the paradigm of Indigenous science has some prerequisites which need to be considered. There must be an understanding and acknowledgment of the history of exploitation of Indigenous peoples by Western culture and science. There must be a willingness on the part of the non-Indigenous teacher to view science from a perspective that is "inside out, upside down and the other way around" or, more simply put, without bias and with deep vision which allows for a deep examination of habitual thought processes. This means reflecting on Indigenous science based on its own terms and methodologies without stereotyping or trivializing its essential components.

The full curriculum model has been formally incorporated in the general Associate of Fine Arts curriculum at the Institute of American Indian Arts in Santa Fe, New Mexico. Adapted forms of the model may be tested at selected contract and public schools serving a predominantly Native American student population. Similarly, forms of the model could be used to introduce Native American students to science related fields through established orientation programs at institutions of higher education. Aspects can be adapted to the Science/Mathematics curriculum development process in schools serving other minority students nationwide.

As mentioned in the preface, the entire curriculum model may be adapted to the science education needs of Indigenous students in other countries. Parallels of the cultural history, needs and orientation which underpin this model also exist among the Maori, Aborigine and tribal African and Asian peoples. Native Hawaiian and Alaska Native Peoples are included in my reference of "Native American." That is to say that this curriculum model can easily be tailored to Native Hawaiian or Alaska Native groups with the substitution of specifically related cultural content.

All subject areas involve experiential exploration of the rational and intuitive-cognitive capacities of students. Therefore, subject matter used to illustrate the process is quite flexible. Because the curriculum is a holistic expression, integration of areas is always possible. Whether it be psychology, ecology, literature and the humanities, language or social science, the model allows for a relevant exploration of the creative process and problem solving skills.

Culture and creativity provide the point of commonality for all types of content. The format of this curriculum model will introduce these elements within most educational contexts in a relatively effective and relevant manner. The model can be applied to other levels (and specific cultural backgrounds) simply by adjusting the content to fit the cultural identity of the students and their relative cognitive/developmental levels.

Conclusions and Recommendations

The approach to science education described within this book presents a significant departure from more conventional approaches. This is because the underlying assumptions are very different when compared to those which have guided curriculum development in the past.

The model, thesis and underlying assumptions have been presented to stimulate the development of new insights for science educators and address the needs of previously uninspired students in the area of science. Many Native American students openly express their negative feelings based on their previous encounters with conventional science curricula. Some even view science as one of the tools Western societies have used to exploit Indigenous populations and natural environments. A student's association of science with exploitation is often a result of a poor understanding of the "science" process within Native American cultures.

As a result of these perceptions and the general lack of attention given to making science culturally relevant, Native American students have tended to reflect their alienation from science with low scores in science and related segments of the Scholastic Aptitude Test and in little interest in entering science related professions. The Council of Energy Resources Tribes based in Denver, Colorado states: "The presence of few Native Americans in science professions places tribes at a decided disadvantage in dealing with resource development and other science-related activities."

In response, this model presents a bridge which allows for an integration of conventional science curricula with a cultural and affective base relevant to Native American students, a first step in reversing the alienation of Native American students toward science.

This curriculum model is based on the assumptions that:

Nature has within it a spirit that is part of each of us. We cannot truly encounter it without changing ourselves, as we affect and are affected by all that we do in relationship to the natural world. This is an essential aspect which infuses meaning into the study of science.

Science education must incorporate both the rational and intuitive thought process.

Science is simultaneously a cultural system of thought, a creative process of problem solving and a system of communication.

Science and art parallel one another as ways of relating to the natural world. Science is an aesthetic experience which involves a wide range of social relationships.

Objectivity is a relative term. Nothing that human beings do can be considered completely objective. Objectivity in the scientific method is the goal of modern scientific endeavors, yet that objectivity is expressed only within the parameters of the paradigm of Western science which itself changes as society does. The reality of relative objectivity is important to convey to students through science curricula.

Science is a form of communication and involves a kind of literacy. This literacy in turn involves the development of basic skill as tools for understanding and solving problems in reference to nature. Such literacy entails an understanding of concepts and natural processes from the perspective of a particular cultural system of thought. It follows from this assumption that science must be approached as a type of dynamic literacy which must be internalized.

How an encounter with natural phenomena affects students and the meaning which it has for them encompasses personal, cultural and creative dimensions of perception. Meaning is the key to relevance. If science is to have meaning for students, that meaning must be inherent in both the content and presentation.

Learning is a natural activity for all human beings. The first step in motivating and enhancing learning of any sort is by encouraging involvement in the learning process. Learning is lifelong and holistic. Modern science education must widen its parameters, open up its paradigm to allow a more holistic and integrated perception of itself to take hold and grow in the minds of students.

The nature of the world today and the projection of the needs of the future require science education which can enable students to develop cognitive abilities more completely. It has been said that the problems which will face the next few generations of mankind will be of a nature and magnitude never before faced in the entire history of man on earth. Science and education will play pivotal roles in terms of how, or even if, these future monumental problems will be solved.

This work is one small step toward a brighter future for Native Americans and Indigenous people everywhere.

In Beauty it is finished...

ACKNOWLEDGMENTS

This work, a synthesis of ideas, experiences and activities, is the result of collaboration with, and inspiration from, a host of people too numerous to mention. I can mention only those individuals who played a major role in helping me bring this work to completion.

First, gratitude must be expressed to the students whom I have had the pleasure to encounter during twenty-five years of teaching Native American students. They have been the major focus and influence for this work. Their needs and their spirit have determined the parameters of the curriculum model which is presented in this book.

Thanks to Dr. Dave Warren and Mr. Jon Wade for their friendship and their support of my research and development of this work during my time with the Institute of American Indian Arts Integrated Studies Project. Special thanks must be extended to Dr. Doug Swartz and the School of American Research in Santa Fe, New Mexico, for their fellowship support during the dissertation phase of this work, and to Dr. Edward T. Hall for his inspiration and special friendship. Dr. Hall's guidance allowed me to explore, create and develop in ways too numerous to recount. Special thanks also to Philip Foss and Sheila Cowing whose excellent editorial work helped to transform my original dissertation into "Igniting the Sparkle".

Finally, very special thanks to my wife Patsy for the patience, understanding, and support which only a wife in her special way can provide... And to my son James whose "sparkling" eyes formed the inspiration for this work twenty years ago and who is now attending the University of New Mexico majoring in computer engineering.

APPENDIX A...

Areas For Further Research And Development

During the process of research and development of this book, a number of possibilities have surfaced.

The creative thought process addresses thought common to all disciplines. Disciplinary areas other than science must be researched in reference to ways that creativity content can be utilized within the implementation of a culturally-based approach.

Many traditional Native American teaching and learning methods contain elements and strategies which parallel many contemporary teaching and learning strategies. Research in these parallels can provide important insights into the process of teaching itself, the transfer of knowledge, and the further development of innovative teaching/learning methods.

The third area for possible further research is the exploration of the effects of bicultural science learning on the enhancement of self-identity of Native American learners, which in turn may lead to increased motivation for pursuit of higher educational goals. Research in this area is of continuing importance if educators are to gain needed insights and strategies for attracting Native American students to and keeping them in college.

The fourth area involves research into the possibilities inherent in the use of "endogenous" cultural knowledge as a basis for integration when teaching and learning another knowledge base. Research along these lines has already been generated through UNESCO's cultural education and development programs over the past decade. However, research dealing with specific applications of endogenous knowledge bases through bicultural education is still needed. This book presents one such attempt along these lines.

The fifth area is further research dealing with the integration of science learning with art learning through their common expression via the creative process. Science and art as cognitive processes, ways of encountering the world and as strategies of problem solving parallel one another in numerous ways. Science tends to emphasize rationalistic-cognitive processes.

Art tends to emphasize intuitive-cognitive processes. Yet both science and art require the integrated application of both the intuitive and rationalistic thought processes during all phases of the creative process. The complementary relationship between science and art is not a new notion. This characteristic of both science and art has significant implications for educating the whole person.

More research might be useful on Native American culture for the purpose of accumulating a body of resources which can be adapted to further bicultural science curriculum development and implementation.

Much has been accomplished through Native American bilingual/bicultural educational research through the implementation of Native American studies in some universities, and as a by-product of research in social sciences like anthropology, history and psychology. However, there is a dearth of translation of this body of research into forms that can be readily used by teachers.

Individual lesson and unit development in bicultural science for Native American students must be a priority. The effectiveness of any curriculum model is directly related to the body of content upon which it relies. If any curriculum is to maintain a consistent balance between the expression of two cultural systems, then content for both systems must reflect a basic parity of development.

Teaching and learning are necessarily intertwined with the act of communication which is multileveled and varies tremendously in relationship to the nature of the situation and subject areas. Research of the situational contexts of communication within the teaching/learning process can have enormous potential for insight in this basic, relatively unexplored dimension of education.

APPENDIX B...

Science From A Native American Perspective Curriculum Syllabi

The following are the actual syllabi of the courses as they are currently taught at the Institute of American Indian Arts in Santa Fe, New Mexico. As such, they outline the requirements of a college level sequence of courses which implement the thesis of "An Indigenous Science Education Model" which I have presented. The elements of these courses can be adjusted for almost any grade level but are more suited for high school or college students. However, by applying the "thematic approach" and selecting or adjusting content for age appropriateness the concepts and themes of these course can be adapted for the upper elementary and mid-school levels as well.

The syllabi presented are but one possibility for organizing the kind of creative and culturally based exploration of science advocated in this work. Each syllabi outline begins with a **"thematic" title** which is chosen based on it descriptiveness of the course content. However, the title can also be a "metaphor" which stimulates thinking and serves as an umbrella for other related sub-themes. The **course philosophy** can be adjusted in numerous ways to fit a particular cultural, disciplinary or instructional approach. In the development of **course goals**, decide on the most important outcomes, standards or learning objectives you hope your students will accomplish and let working toward the realization of your goals guide implementation and processing of the course. The **course of study** can revolve around key questions, key concepts, experiences or sub-themes which you wish to explore. This segment denotes the most important content areas you want to cover in the implementation of the course. The **evaluations** should be specific yet flexible in the ways in which students may demonstrate their learning. The use of portfolio type evaluations, art projects or social action activities are ideal. The key is to give students diverse criteria for evaluations which go beyond "paper and pencil testing." The final section of **suggested reading** can include a number of different selections which represent different aspects of the course theme or otherwise enhance students learning. I have given a few examples of readings I have used, however there are literally hundreds of possible books, articles, reference and triballly specific curricular materials which can be used by both students and teacher to provide reflective reading and content for these courses. **The key is to be creative in your development process. Use the following to reflect upon but dare to create your own syllabi! Remember that curriculum development is a creative process and a function of design and apply these principles throughout.**

SYLLABUS #1...
The Creative Process in Art, Science and Native American Cultures

COURSE PHILOSOPHY:

This course is an exploration of the dynamic and holistic nature of the creative process and its reflection in art, science and Native American cultures. Special attention will be paid to inquiry, abductive reasoning, experience, reflective dialogue and application as the basic modes of learning. Areas to be explored with reference to Native American cultural and social perspectives will include creative problem solving in applied science, the meditative mind, visualization, creative movement, creative writing, film, artistic symbolization, music, dance and drama. The main goal of this course is to give the student the opportunity to develop a basic grounding in the creative process as it is reflected in the natural world and in the context of Native American cultural expression. In addition, students will be given the opportunity to develop insights into their own unique creative potential and cultural selves.

COURSE OBJECTIVES:

A. To gain a basic understanding of what "creativity" is—its basic characteristics and processes, its verbal and non-verbal and its conscious and unconscious forms and its expressions in the natural world.

B. To develop personal methods or approaches to creative synthesis based on basic understanding of self and one's own unique creative capacities.

C. To develop perspectives of the creative thought process in art, science and Native American culture.

D. To develop a working knowledge of the application of inquiry, reason, intuition, visualization, meditation, creative problem solving and learning.

COURSE OF STUDY: An Inquiry or the Art of Asking the Right Question.

Phase One: An overview of the creative process.

-What is the nature of the creativity? How does it reflect the characteristic thought processes of the right and left hemispheres of the brain? How is creativity related to the innate human instinct to relate to other living things called "biophilia"?

-How does creativity evolve and transform through time? What are the basic phases of the creativity process? How are they reflected in art? In science? In Native American cultures? What is the predominate image of creativity? How is creativity inherently tied to the creation of images?

-What roles do reason, intuition, abduction, visualization, inquiry and "structure" play in the creative thought process?

-How is the creative mind reflected in art, in science, in healing, in religion, in language, in culture?

-The creative body - in art, science, in healing, in dance, in drama, in culture

-What about creativity and evolution, creativity and the forms of nature, creativity and the cosmos, creativity and the archetypal elements?

-How is creativity enhanced through meditational art, dance, music, drama, poetry, writing, teaching, healing, play and ceremony?

Phase Two: Creativity in Native American Traditions

-How is the nature of creativity portrayed in Native American creation myths, origin stories and trickster tales?

-What about the metaphoric personifications of the four cardinal directions of the creative mind: the warrior, the shaman, the artist, the philosopher?

-What is the "ceremony of art" as it is exemplified in traditional Native American art forms?

-How do Native American Cosmologies compare with those of Western and Eastern traditions?

-What are the visionary and artistic foundations of tribal education?

-What role does the symbol and metaphor of the "Tree of Life" play in Native American conceptualization of the nature of creativity?

-How is the nature human expression "Biophilia" played out in Native American orientations to creativity?

-How is creativity culturally defined in Native American tribal communities?

TEACHING METHODS:

The basic teaching strategies to be used in this class will be an adapted form of the creative process teaching/learning round, inquiry and experiential modes of instruction using primarily an encounter- reflection-discussion-demonstration-experiential activity and presentation format. Primary emphasis will be given to the application of the holistic thought processes of the mind in both the inquiry and experiential phases of classroom interaction.

Basic research, experiential exercises, field trips, guest speakers and film review will be used periodically to attain course objectives.

EVALUATION:

1. Evaluation of progress made by each student in developing a working understanding of the creative thought process as demonstrated through a personal portfolio of the creative insight and integration of course experiences

2. Evaluation of students demonstrating how he or she is applying their understanding of Indigenous creativity to a unique project or art form of their choice.
3. A final self-evaluation, with guidance from instructor, or each individual students growth and development in the application or expression of the creative process.
4. The quality and "nature" of class participation.
5. A basic research paper in an area of "The Creative Process."

SUGGESTED READING:

Steps to an Ecology of the Mind by Gregory Bateson
Drawing from the Right Side of the Brain by Betty Edwards
Human Teaching and Human Learning by George I. Brown
Education through Art by Herbert Read
Teaching Creativity through Metaphor by Donald A. Sanders
Look to the Mountain by Gregory A. Cajete
Science, Order and Creativity by David Bohm and F. David Peat
The Metaphoric Mind by Bob Samples
In the Absence of the Sacred by Jerry Mander
Metapatterns: Across Space, Time and Mind by Tyler Volk
The Spell of the Sensuous by David Abram
Seven Lessons of Chaos by John Briggs and F. David Peat
The Primal Mind by Jamake Highwater
The Universe Story by Brian Swimme and Thomas Berry
The Mission of Art by Alex Grey

SYLLABUS #2...
Philosophy:
A Native American Perspective

COURSE PHILOSOPHY:

This course in its essence is survey of the "ordering" philosophical paradigms of Native North, Central and South America. Its emphasis will be an exploration of the ways in which these "ordering paradigms" have guided the thoughts, values, aesthetics, ethics and actions of Native American expressions of science, art forms, artistic symbols, oral poetry, mythology and literature. It is hoped that such a broad examination will stimulate further interest in these areas and form a resource which will enhance individual creativity.

COURSE OBJECTIVES:

a. To develop conscious awareness of the major "ordering" philosophical paradigms of Native North, Central and South America.
b. To develop an understanding of the ways these ordering paradigms have guided Native American worldview and philosophy.
c. To gain a basic perception of the nature of philosophy, its characteristic forms of inquiry and expression in cultural arts and sciences.
d. To gain insight into the ways in which Native American philosophical thought is reflected in Native American ethnosciences, art, song, music, dance, community, poetry, mythology and social ecology.

COURSE OF STUDY:

Inquiry and the Art of asking the right question.

-What is cultural philosophy? Why it considered the foundation of all arts and sciences?

-How does philosophy reflect the right and left brain thinking processes?

-What is a paradigm? What are the major "ordering paradigms" of Native America? The life symbols? The tree of life, the sacred twins, the supreme wind, the great mystery ?

-How have these ordering paradigms influenced the aesthetics, and basic values and philosophy of Native American cultures?

-What is the inherent power of language? What are the roots of myth?

-Nature and the "primal' philosophy of Native America? Natural philosophy?

-Life symbols and the roots of sacrament? An exploration of the Native American mindset of "seeking life" as reflected in Native American traditional philosophy.

-How are "designs" of the natural world reflected in Native American aesthetics?

-How are matter and energy and their implicate and explicate order as reflected in Native American and Western philosophical perspectives of the transformations of energy?

-Space and time - A Native American perspective and Quantum physics? How do these different perspectives affect the worldview of cultures?

-Modern systems theory and Native American traditional philosophy.

-The nature of things and the things of nature? A parable. The natural world and its influence on Native American philosophical perspectives.

-Image + Identity = Image making. Forms of Native American artistic expression and Native American philosophical expressions of natural phenomena and relationship.

-The sensitization of the "seven senses" - touch, smell, taste, hearing, sight, intuition, and spirituality? A genesis of artistic importance

- the Native American approach.

-Primal man - Philosopher par excellence! Philosophy and social behavior.

-A comparison of Native American cosmologies.

TEACHING METHODS:

The basic strategy to be used in this class will be adapted form of the creative process teaching/learning round, inquiry and experiential modes of instruction using primarily an encounter- reflection-discussion-demonstration-experiential activity and presentation format. An inquiry process which incorporates abductive modes of research and synthesis will be encouraged. Great emphasis will be placed upon the "art of forming paradigms" based on the observation of natural processes and phenomena. The creative thought process and its reflection in paradigm formation which generate philosophical perspectives in the arts and sciences will also be explored. Basic research, guest speakers and film review will also be an integral part of this course.

EVALUATION:

1. The "quality" and nature of class participation.
2. A cumulative take home final examination.
3. A basic research paper in an area of "Philosophy: from a Native American Perspective."
4. Understanding and skilled application of the "philosophical" thought process as demonstrated through the creation and completion of an appropriate project.

SUGGESTED READING:

Philosophy in a New Key by Susanne K. Langer
The Primal Mind by Jamake Highwater
The Structure of Scientific Revolutions by Thomas S. Kuhn
The Essential Tension: Selected Studies in Scientific Tradition and Change by Thomas Kuhn
Science and Subjectivity by Israel Scheffler
Primitive Man as Philosopher by Paul Radin

The Old Ways by Gary Snyder
The Sacred: Ways of Knowledge Sources of Life by Peggy Beck and Anna Walters
Topophilia by YI-fu Tuan
Indian Ecology by Donald Hughes
General Systems Theory by Ludwig Von Bethel
Look to the Mountain by Gregory A. Cajete
A Sacred Place to Dwell by Henryk Skolimowski
The Participatory Mind by Henryk Skolimowski
The Dream of the Earth by Thomas Berry
The Primal Mind by Jamake Highwater
Seven Lessons of Chaos by John Briggs and F. David Peat
Wisdom of the Elders by David Suzuki and Peter Knudtson
Sacred Land, Sacred Sex by Dolores LaChapelle

SYLLABUS #3...
Social Ecology: A Native American Perspective

COURSE PHILOSOPHY:

The main intent of this course will be to help each student gain a greater awareness and understanding of the interdependence of the social and ecological paradigm and their resultant influences on the cultural psychology and social ecology of Native American people, past and present. Through such insights it is hoped that each student will gain an understanding of the social dynamics of Native American tribalism. Developing a basic understanding of social psychology using Post-colonial and Jungian perspectives will be an integral part of this course.

COURSE OBJECTIVES:

1. To develop a basic understanding of the conscious and unconscious reflections and process of social ecology in the emergence and creation myths of selected Native American tribes.

2. To gain a greater awareness of how the "guiding paradigms" of Native American mythology have influenced the behavior, ethics and social psychology of Native American society.

3. To gain a greater awareness and understanding of the dynamics of the social, psychological and ecological forces at work in the indigenous world today. Through this basic awareness each student will have some of the insights needed to cope with these forces in a constructive way.

4. To gain empathy for the uniquely Native American value structures and the psycho-spiritual orientations of the dynamic of the communal mind.

5. To become aware of the way in which Native Americans histori-

cally applied the characteristic thought process of "ecology" in combination with the artist perception and the philosopher's synthesis.

6. To gain an understanding of the general concepts of Social Ecology and some of its applications.

COURSE OF STUDY: An Inquiry

What are the basic concepts of Social Ecology? In what ways has man used these concepts to order and understand his existence? What is Post-Colonial Psychology? What are its basic precepts?

Who was Carl Jung? What are some of his ideas about social psychology and the unconscious mind? How do the concepts and perceptions of tribal relational philosophy and Carl Jung relate to and help interpret the social psychology of Native America?

Native American Creation myths as a reflection of the social and psychological dynamics of the "indigenous" mind.

Ethics, moral and amoral behavior: The trickster friend or foe buffoon or sage?

The Earth mother/Sky father, the Sacred Twins, Sister Moon, the Culture Hero, Spider woman, the Holy Winds, Great Serpents, the Thunderbird, the Cloud Being, the Corn Mother, Archetypes and the Native American Collective Unconscious?

Within the context of Native American tribal and community life, What value structures and practices enhance integration of personality and individuation?

How does the traditional social organization of Native American Communities influence the physical, psychological and spiritual health? What factors are involved which determine the directions this influence will take? What are some ways to ensure the influences will be positive?

What is the relationship between Acculturational psychology and Indian culture?
What are the major Traditional guiding values of Native American

cultures? How did they arise? How are they faring In Indian life today? What is the "wind standing within and the "wind standing without"?

How does one creatively deal with Cultural conflict and alienation? What are some specific conflicts of Indian and non-Indian values and how might one deal with them?

What is mental health and wholeness from a Native American Perspective? What role does Shamanism play in terms of this perspective?

What does the future hold for the changing and ever evolving social and ecological personality of Native America?

TEACHING METHODS

The basic teaching strategies to be used in this class will be a modified social inquiry method of instruction using primarily the lecture-discussion and seminar format of presentation. Special emphasis will be placed upon pertinent information and current research relating to concepts in Social Psychology, Acculturational Psychology, Mythology and holistic mental health which is relevant to Native American life. Also, emphasis will be placed upon the social research and participant aspects of individual basic research, guest speakers, field trips and film review will also be used periodically to attain the stated course objectives learning.

EVALUATION:

1. A cumulative take home examination.
2. Regular class attendance
3. The quality and nature of class participation
4. A basic research paper in an area of "Social Psychology: From a Native American Perspective."
5. Understanding and skilled application of the creative
thought process as demonstrated through the completion and presentation of an appropriate project.

SUGGESTED READING:

Jung, His Life and Work by Barbara Hannah
Introduction to Humanistic Psychology by Charlotte Buhler and

Melanie Allen
A Magic Dwells by Sheila Moon
The Sacred: Way of Knowledge Sources of Life by Peggy Beck and Anna Walters
Modern Indian Psychology by John F. Bryde
Indians of Today by Marion E. Gridley
Speaking of Indians by Ella Deloria
Health and Wholeness by John Sanford
Post-Colonial Indian Psychology by Eduardo Duran and Bonnie Duran
As We See: Aboriginal Pedagogy Ed. by Lenore Stiffarm
Look to the Mountain by Gregory A. Cajete
Wisdom of the Elders by David Suzuki and Peter Knudtson

SYLLABUS #4...
Herbs, Health and Wholeness: A Native American Perspective

COURSE PHILOSOPHY:

This class will revolve around various aspects of herbalogy and concepts of holistic health in Native American. Its main intent will be to introduce students to the history, actual practical application and underlying concepts of health and wholeness which govern the use of herbs and other therapeutic methods beginning with traditional Native America perspectives and then working through to traditional Chinese, Ayurvedic and contemporary holistic health therapeutic practices and perspectives.

COURSE OBJECTIVES:

a. To gain a basic understanding of the development, roles and influences of medicinal plants and holistic health concepts upon the worldview and realities of Native America past and present.

b. To gain a greater awareness of the natural environment and the ecology of the plants therein.

c. To gain a basic working understanding of identification, methods of gathering and preparing and actual therapeutic uses of plants by Native Americans past and present.

d. To gain an appreciation for the contribution made by Native America in the areas of pharmacology, medical therapeutics and Agriculture.

e. To develop an empathy based on understanding of the psycho/spiritual and physical natures of illness, health, and wholeness from a Native American perspective.

f. To gain basic insights into the general principles of holistic health and its maintenance.

g. To become aware of the nature of the "healing arts" through which Native Americans applied the characteristic thought process of "science" in combination with the artist's perception and the philosopher's synthesis.

h. To gain an empathy for the uniquely Native American value structures and psycho/spiritual interpretations of the dynamic natures of human and plant ecology and the web of life.

COURSE OF STUDY:

-What is herbalogy? What are its origins and influences? What is the extent of the practice of herbalogy and other alternative healing systems in the U.S.? In the world?

-Native American concepts of illness, health and wholeness - An adaptation to the natural environment?

-How has Native American health changed since the time of Columbus? What are some of the reasons for these changes?

-What is the characteristic nature of Native American healing arts? What are some of the principle concepts by which it is practiced?

-What is Shamanism? What role does Shaman play in Native American healing?

-The concept of medicine from a Native American perspective - what is it? In what ways is it expressed?

-What are the characteristics of the life histories of Medicine people?

-What are some Native American methods of diagnosis and treatment of disease?

-What roles have mind altering plants played in Native American healing and why? Chinese and Ayuruedic medicine?

-What are some of the most common valid medicinal and edible plants used by Native Americans? How are they identified?. How are they prepared? How are they used?

-How are herbs used in traditional New Mexico? What is the "history" of this herbalism of New Mexico? How does this history reflect the dynamics of adaptation and evolution or herbalism in other regions of America?

-In what way does herbalism reflect the general principles of holistic health practices? What are the basic principles of herbology? How do these principles and practices reflect systems ecology and systems theory? What is a system?

-How do Eastern healing systems such as Chinese medicine or Ayuruedic medicine compare to Native American healing philosophies and practices?

-What are the basic methods of gathering, preparing, and storing herbal medicines?

TEACHING METHODS:

The basic teaching strategies to be used in this class will be a modified combination of the inquiry and experiential modes of instruction using primarily an encounter- reflection-discussion-demonstration-experiential activity and presentation format. Special emphasis will be placed upon pertinent information and current research relating to concepts in holistic health, herbalogy, psychology, philosophy and systems ecology from a Native American perspective.

Also, the application of the creative dimensions of the "scientific and artistic" thought processes in both the transfer and inquiry phases of instruction will be an integral part of this course.

Basic research, labs, guest speakers, film review and field trips will be used periodically to attain stated course objectives.

EVALUATIONS:

1. The "quality" and nature of class participation.
2. A cumulative take home final.

3. A basic research paper in the area of herbs, health and wholeness from a Native American perspective.

4. Understanding and skilled application of the "creative science/art thought process" as demonstrated through the completion and presentation of an appropriate project.

SUGGESTED READING:

American Indian Medicine by Virgil Vogel
Indian Herbology by Alan Hutchinson
Dimensions of Holistic Health Education by Otto & Knight
Healing Herbs of the Upper Rio Grande by L.S.M. Curtin
Medicinal Plants of the Mountain West by Michael Moore
Healing and Wholeness by John Sanford
The Shaman by Joan Halifax
American Indian Healing Arts by E. Barrie Kauasch and Karen Baar
Cultures of Habitat by Gary Paul Nabhan
A People's Ecology by Gregory A. Cajete
In the Three Sisters Garden by Joanne Dennee, Jack Peduzzi and Julie Hand
Sacred Trees by Nathaniel Altman
Look to the Mountain by Gregory A. Cajete
Keepers of the Plants by Michael J. Caduto and Joseph Bruchac

SYLLABUS #5...
Animals in Native American Myth and Reality

COURSE PHILOSOPHY:

An exploration of Tribal/Indigenous education, ecology and philosophy as seen from the perspective of animal mythology and actual relationship to animals. Emphasis will be placed on this special relationship as expressed in art, applied "science", visionary experiences and environmental ethics. This integrated and holistic exploration will form the foundation for a more complete understanding and appreciation of mythology and tribal indigenous educational processes.

COURSE OBJECTIVES:

a. To develop perspectives through the lens of mythology of the role animals have play in Native American life as expressed through song, dance, art and ritual play in tribal/indigenous education.

b. To develop basic understandings of the concepts and applications of mythological thinking.

c. To become aware of the ways tribal Americans apply the characteristic thought processes inherent in myth in combination with the artist's perception and the philosopher's synthesis to develop a truly holistic context for teaching and learning about animals and human relationships therein.

COURSE OF STUDY:

a. The nature of myth and its connection to teaching and learning.

b. An alternative analysis of what we now call Indian education.

c. The seven directions of tribal/indigenous knowledge as perceived through Native American mythology about animals.

d. The Mythic Foundation of tribal/indigenous relationship to animals.

e. The Environmental Foundation of tribal/indigenous relationship to animals.

f. The Artistic Foundation of tribal/indigenous relationship to animals.

g. The Visionary Foundation of tribal/indigenous relationship to animals.

h. The Community Foundation of tribal/indigenous relationship to animals.

i. The Affective Foundation of tribal/indigenous relationship to animals.

j. The Spiritual context of tribal knowledge and relationship to animals.

k. The challenge of revitalizing tribal education and relationship to animals in the context of the contemporary educational forms of the 21st Century.

TEACHING METHODS:

The basic teaching strategies to be used in this class will be a modified combination of the Creative Process Teaching/Learning Round, ITI (Integrated Thematic Instruction) and Inquiry/Dialogue modes of instruction using primarily the lecture, discussion and seminar formats of presentation. Special emphasis will be placed upon pertinent information and current research relating to concepts in Tribal education, Native American philosophy and Mythology. Also the application of the creative dimensions of the 'scientific and artistic" thought processes in both the transfer and inquiry phases of instruction will be an integral part of this course.

Basic research, guest speakers, film review and field trips will be used periodically to attain the stated course objectives.

EVALUATIONS:

Evaluations will be based on the following criteria of performance:

1. A portfolio documenting evolution of personal insight through the course.

2. A cumulative take home examination.

4. The "quality" and nature of class participation.

5. A basic research paper in an area of "Myth and Education from a Native American Perspective."

6. Understanding and skilled application of the creative /mythic/art thought process as demonstrated through completion and presentation of an appropriate project.

SUGGESTED READING:

American Indian Mythology by Alice Marriott and Carol Machlin
Black Elk Speaks by John G. Neihardt
Pueblo Animals and Myth by P.C. Tyler
Pueblo Birds and Myth by P.C. Tyler
The Old Ways by Gary Snyder
Myth and Reality by Mircea Elliade
The Way of Animals by Joseph Campbell
The Sacred: Way of Knowledge, Sources of Life by Peggy Beck and Anna Walters
The Inner Reaches of Outer Space: Metaphor As Myth Religion by Joseph Campbell
American Indian Myths and Legends by Alfonso Ortiz and Richard Erdoes
Animals of the Soul by Joseph Epes Brown
The Spell of the Sensuous by David Abram
Keepers of the Animals by Michael J. Caduto and Joseph Bruchac

SYLLABUS #6...
The Primal Elements:
A Native American Perspective

COURSE PHILOSOPHY:

This course will explore the Native American ecological ethic as expressed in the representations of thought and action with regard to the "primal" elements of the physical world. The format of the course will revolve around an exploration of the ecological relationships and understandings of these elemental forces of nature as they are reflected in the "cultural fabric" of Native America. The areas that will be emphasized include art, architecture, philosophy, and science. A basic but very general understanding of physical geo-science will be an integral part of this course.

COURSE OBJECTIVES:

a. To gain a greater awareness of the natural environment and man's relationship to the elemental forces therein.

b. To develop perspectives of the roles which the archetypal elements of nature have played in the natural philosophy, arts, mythology, and human ecology of Native America.

c. To develop basic understandings of the concepts and applications of environmental geo-science and the characteristic essences of "ether, air, fire, water, and earth".

d. To become aware of the ways in which Native Americans historically applied the characteristic thought process of "science" in combination with the artist's perception and the philosopher's synthesis to develop a truly holistic conception of geo-science and the archetypal elements of nature.

COURSE OF STUDY: An Inquiry

-What are the Archetypal Elements? How have they been viewed by the civilization of man?

-How has an understanding of the archetypal elements influenced such things as philosophy, medicine, art, architecture, mythology, and religion? Worldwide? In Native America?

-What are the essential characteristics of ether? What are its general interpretations from a Native American point of view? What are its basic manifestations in the Cosmos, in nature, and in man? In what ways is it symbolized in the various aspects of Native American cultures? Art, architecture, and myths?

-What are the essential characteristics of air? What are its general interpretations from a Native American point of view? What are its basic manifestations in cosmos, in nature, and in man? In what ways is it symbolized in the various aspects of Native American cultures? What is the place of the holy wind in Native American philosophy?

-What are the essential characteristics of fire? What are its general interpretations from a Native American point of view? What are its basic manifestations in the cosmos, in nature, and in man? In what ways is it symbolized in the various aspects of Native American culture? Art, architecture, and myth?

- What are the essential characteristics of water? What are its general interpretations from a Native American point of view? What are its basic manifestations in the cosmos, in nature, and in man? In what ways is it symbolized in the various aspects of Native American cultures? Art, architecture, and myth?

- What are the essential characteristics of earth? What are its general interpretations from a Native American point of view? What are its basic manifestations in the cosmos, in nature, and in man? In what ways is it symbolized in the various aspects of Native American cultures? Art, architecture, and myth?

- Are the archetypal elements and their dynamic interacting natures the architects of the universe ? How is their cosmic activity reflected on Earth? Its life processes, its weather, its movements, its internal activities?

- How did ancient peoples study of the activities of the archetypal elements? In what ways is this study reflective of Native American philosophies? What are the general principles of modern geo-science?

- The cosmos is the textbook - seek from it, learn from it. Nature is the text book seek from her.

- Man is the textbook seek from him, learn from him. Some thoughts to think about.

TEACHING METHODS:

The basic teaching methods to be used in this class will be a modified combination of the creative process teaching/learning round and inquiry modes of instruction using primarily the lecture-discussion-demonstration and seminar format of presentation. Special emphasis will be placed upon pertinent information and current research relating to concepts in environmental geo-science, physics, medicine, philosophy, and environmental physiology, and ecology.

The application of the creative dimensions of the science/art thought processes in both the transfer and inquiry phases of instruction will be an integral part of this course.

Basic research, labs, guest speakers, film review, and field trips will be used periodically to attain stated course objectives.

EVALUATIONS:

1. A portfolio documenting evolution of personal insight through the course.

2. The "quality" and nature of class participation.

1. A cumulative take home examination.

4. A basic research paper concerning some aspect of one of the archetypal. elements as viewed from a Native American perspective.

5. Understanding and skilled application of the "creative science/art" process as demonstrated through completion and presentation of an appropriate project.

SUGGESTED READING:

Sensitive Chaos by Theodore Schwenk
The Seven Mysteries of Life by Guy Murchie
The Earth We Live On by Ruth Monroe
Wisdom of the Elders by David Suzuki and Peter Knudtson
The Sacred Balance by David Suzuki
Metapatterns: Across Space, Time and Mind by Tyler Volk
The Way of the Earth by John Bierhorst
A Sacred Place to Dwell by Henryk Skolimowski
The Universe Story by Brian Swimme and Thomas Berry
Keepers of the Earth by Michael J. Caduto and Joseph Bruchac

SYLLABUS #7...
Astronomy: A Native American Perspective

COURSE PHILOSOPHY:

This course is designed to provide students with appropriate content, learning experiences and other resources related to the cultural expression of astronomy in Native America. The aim of such an exploration is to engender in participants a well contexted understanding and appreciation of the skills, approaches, cultural perspectives and guiding cosmologies which characterized Native American expressions of astronomy. Special emphasis will be placed on facilitating the integration of the exploration of basic astronomy with a basic understanding of the influences the observation of the stars, planets, moon, sun and other celestial phenomena have had on the cultural perceptions of Native Americans, past and present.

OBJECTIVES:

1. To develop a basic understanding of the influences of ancient Native American observations of the heavens upon the cultural world views and cosmologies of Native America.

2. To gain a greater awareness of the cyclic and dynamic motion of the visible celestial bodies and phenomena.

3. To develop skill in the transfer of knowledge about Native American perceptions of astronomy to other areas of knowledge and creative expression in the arts.

4. To gain basic appreciation for the uniquely Native American psycho/spiritual interpretations of the ebb and flow of the cosmos.

5. To become aware of the way in which Native Americans historically applied the characteristic (cultural) thought process of Science in combination with the artist's perception and the philosopher's synthesis.

COURSE OF STUDY:

PHASE ONE: PRELIMINARIES

-How does the Universe operate?

-What are the Basic Cycles of Celestial Movement? The Sun, Moon, Venus, Mercury, Pleaides, day/night, the Seasons and the Year?

-What about the evidence of ancient celestial observations? What do they tell us about man's long watch of the heavens? What do they tell us about the origin of cosmological ideas?

-What are the reflections of the Sacred in ancient celestial knowledge? How are these expressed in concepts of Divinity, Worship and Ritual-Mother Earth and Father Sky in Native American traditions.

PHASE TWO:

- What the guiding concepts of sky cycles and cosmic myths? The reflection of cosmic order in Native American mythology.-What are the ancients' guiding concepts of afterlife and their cosmic connections? Death, dying, dreaming, rebirth, and the celestial journey - A Native American perspective.

-What role did the ancient astronomers play in their ancient societies? Shamans, Star Priests, Moon Priests, Sun Priests, Ancient Astrologers and Astronomers.

-What about the Sacred dimension of Time? The role of the Great Calenders of the Americas.

-What about the Sacred dimension of Space? The reflection of sacred space of the cosmos in Native American life and traditions.

PHASE THREE:

-What characterized the celebrations of the Sky? The cyclic ceremonies of Native America.

-How did astronomical knowledge characterize Native American Architecture- a Microcosm of the Macrocosm? The Cosmic symbolism and orientations of the scared structures of Native American cultures.

-How was astronomical knowledge represented in Native American Sacred Art? The visionary and psycho/spiritual reflection of the Cosmos in Native American traditional arts.

TEACHING METHODS:

The basic teaching strategies to be used in this class will be a modified combination of the Creative Process Teaching/Learning Round, ITI (Integrated Thematic Instruction) and Inquiry/Dialogue modes of instruction using primarily the lecture discussion demonstration experiential format. Special emphasis will be given to the application of the creative dimensions of the Science thought process in both the transfer and inquiry phases of classroom interactions. Also, emphasis will be placed upon the experiential and hypothesis formation and testing aspects of individual learning. Basic research, guest speakers, field trips, lab experience and film review will also be used periodically to attain the stated course objectives.

EVALUATION:

1. A portfolio documenting evolution of personal insight through the course.

2. The quality of class participation.

3. A creative research project in an area of "Astronomy: a Native American Perspective."

4. An artistic creation which reflects a creative reflection of student understanding of Native American Astronomy.

SUGGESTED READING:

Cosmos by Carl Sagan
Beneath the Moon and Under the Sun by Tony Shearer
Native American Astronomy by Anthony F. Aveni
Astronomy: The Evolving Universe by Michael Zeilik
Living the Sky by Fred Williamson
Earth is My Mother, Sky is My Father by Trudy Griffin-Pierce
Ancient Astronomers by Anthony Aveni
Stars of the First People by Dorcas S. Miller
Lines to the Mountain Gods by Evan Hannigham
Native American Mathematics Ed. by Michael P. Closs
The Universe Story by Brian Swimme and Thomas Berry
The Mind of God by Paul Davies
A Beginner's Guide to Constructing the Universe by Michael S. Schneider
Maya Cosmogenesis by John Major Jenkins

Comprehensive Bibliography

- Aamodt, Agnes Marie (1971). *Enculturation Process of the Papago Child. An Inquiry in the Acquisition of Perspective on Health and Healing.* Ph.D. Dissertation. Tucson: University of Arizona.
- Abruscato, J. (1996). *Teaching Children Science.* Boston: Allyn and Bacon.
- Adams, D.W. (1990). Fundamental considerations: The deep meaning of Native American schooling. 1880-1900. *Havard Educational Review,* 58(1), 1-28.
- Aikenhead, Glen S. (1997). Toward a First Nations Cross-Cultural Science and Technology Curriculum. *Science Education.* Vol. 81.
- *American Indian education handbook.* (1982) American Indian Education Unit California State Department of Education.
- Anders, P.L., & Lloyd, C.V. (1989). The significance of prior knowledge in the learning of new content-specific instruction. In D. Lapp, J. Flood, & N. Farnan (Eds.). *Content area reading and learning: Instructional strategies* (pp. 258-271). Englewood Cliffs, NJ: Prentice Hall.
- Atwater, M.M. (1993). Multicultural science education: Assumptions and alternative views. *The Science Teacher,* 60(3), 32-37.
- Alder, Peter S. (1974). Beyond Cultural Identity: Reflections on Cultural and Multicultural Man. pp. 23-40. *Topics in Culture Learning.* Richard W. Brislin Ed. Vol. 2, 1974. Honolulu: East-West Center Culture Learning Institute.
- Au, K.H. (1990). Participation structures in a reading lesson with Hawaiian children: Analysis of a culturally appropriate instructional event. *Anthropology and Education Quarterly,* 11(2), 91-115.
- Aurbach, Herbert A., ed. (1967). *Proceedings of the National Research Conference on American Indian Education.* Kalamazoo: Society for the Study of Social Problems.
- Aurbach, Herbert A. and Estelle Fuchs with Gordon MacGregor (1970). *The Status of American Indian Education.* An interim report of the National Study of American Indian Education to the Office of Education, U.S. Department of Health, Education, and Welfare Project No. 8-0147. The Pennsylvania State University.
- Banks, James A. (1973). *Teaching Strategies for the Social Studies.* Reading: Addison-Wesley Pub.
- Barba, R.H. (1995). *Science in the Multicultural Classroom.* Boston: Allyn and Bacon.

- Barber, Bernard and Walter Hiroch (Eds.) (1972). *The Sociology of Science*. New York: The Free Press of Glencoe.
- Barnes, Barry (1982). *T.S. Kuhn and Social Science*. New York: Columbia University Press.
- Basso, Keith H. and Henry A. Selby (1976). *Meaning in Anthropology*. Albuquerque: School of American Research/University of New Mexico Press.
- Battiste, M. and J. Barman (Eds.) (1995). *First Nations Education in Canada: The Circle Unfolds*. Vancouver: University of British Columbia Press.
- Beck, P.V. and A.L. Walters (1977). *The Sacred: Ways of Knowledge, Sources of Life*. Tsaile, AZ: Navajo Community College Press.
- Black, Mary Bartholomew (1967). *An Ethnoscience Investigation of Ojibwa Ontology and World View*. Ph.D. Dissertation, Sanford University.
- Blakeslee, Robert (1980). *The Right Brain*. Garden City: Anchor Press/Doubleday.
- Bohm, David (1983). *Wholeness and the Implicate Order*. London: Ark Paperbacks.
- Brinkman, William W. & Leherer, Stanley (1972). *Education and the Many Faces of the Disadvantaged*. New York: John Wiley & Sons.
- Briskman, Larry (1981). *Creative Product and Creative Process in Science and Art* in *The Concept of Creativity in Science and Art*. Dutton, D. and M. Kransz (Eds.) The Hague: Martinus Nifhoff Pub.
- Brown, Joseph Epes (1982). *The Spiritual Legacy of the American Indian*. New York: Crossroads Pub.
- Bruner, J. (1960). *The Process of Education*. Cambridge: Harvard University Press.
- Bryant, Billy Joe (1974). *Issues of Art Education for American Indians in B.I.A. Schools*. Ph.D. Dissertation: Penn State.
- Bryde, John F. (1970). *The Indian Student, A Study of Scholastic Failure and Personality Conflict*. Vermillion: The Dakota Press.
- Burger, Henry (1968). *Ethno-Pedagogy*. Albuquerque: Southwestern Cooperative Educational Laboratory.
- Cajete, Gregory A. (1986) *Science a Native American Perspective: A Culturally-Based Science Education Curriculum*. Ph.D. Dissertation. Los Angeles: International College.
- Cajete, Gregory A. (1994). *Look to the Mountain: An Ecology of Indigenous Education*. Durango: Kivaki Press.
- Cajete, Gregory A. (1999). The Making of an Indigenous Teacher. (161-183) In Kane, Jeffery (Ed.). *Education, Information and Transformation: Essays on Learning and Thinking*. Upper Saddle River: Prentice-Hall.

- Cajete, Gregory A. (1999). *A People's Ecology: Explorations in Sustainable Living.* Santa Fe: Clearlight Publishers.
- Cajete, Gregory A. (1999). Reclaiming Biophilia: Lessons from Indigenous Peoples. (189-206). In Smith, G.A. and D.R. Willaims (Eds.). *Ecological Education in Action.* Albany: State University of New York Press.
- Campbell, Joseph (1983). *The Way of Animal Powers.* London: Summerhill Press, San Francisco: Harper & Row.
- Capra, Fritjof (1976). *The Tao of Physics.* Boulder: Shambhala Publications.
- Capra, Fritjof (1982). *Turning Point.* New York: Simon & Schuster Publications.
- Cave, William M. and Chesler, Mark A. (1974). *Sociology of Education.* New York: MacMillan Publishing Co.
- Clarkson, P.C. (1991). *Bilingualism and mathematics learning.* Geelong, Victoria, Australia: Deakin University Press.
- Cleary, L. and T. Peacock. (1998) *Collected Wisdom.* Boston: Allyn and Bacon.
- Cobern, W.W. (1991). *Contextual constructivism: The impact of culture on the learning and teaching of science.* Paper presented at the Annual Meeting of the National Association for Research in Science Teaching (Lake Geneva, WI., April 7-10).
- Cole, Michael, John Fay, Joseph Glick, and Donald Sharp (1971). *The Cultural Context of Learning and Thinking.* New York: Basic Books, Inc.
- Cole, M. & Griffin, P. (1987). *Improving science and mathematics education for minorities and women: Contextual factors in education.* Madison: Wisconsin Center for Education Research.
- Cortes, C.E. (1986). The education of language minority students: A contextual interaction model. In C.F. Leyba (Ed.), *Beyond language: Social and cultural factors in schooling language minority children* (pp. 3-33). Los Angeles: Evaluation, Dissemination and, and Assessment Center California State University, Los Angeles.
- Cole, M. & Scribner, S. (1974). *Culture and Thought: A Psychological Introduction.* New York: Wiley Press.
- Coomaraswamy, Ananda (1962). *Hinduism and Buddhism.* Westport: Greenwood Press (6- 23, note 21).
- Costa, V.B. (1995). When science is "another world:" Relationships between family, friends, schools, and science. *Science Education,* 79(3), 313-333.

- Driver, R., & Oldham, V. (1986). A constructivist approach to curriculum development in science. *Studies in Science Education*, 13, 105-122.
- Deloria, Vine (1973). *God is Red.* New York: Dell Publishing.
- Dennis, Wayne (1940). *The Hopi Child.* New York: Appleton Century Co.
- Dunn, R. (1983). Learning styles and its relation to exceptionality at both ends of the spectrum. *Exceptional Children*, Vol. 49, No. 6.
- Dutton, Dennis and Michael Krausz, ed. (1981). *The Concept of Creativity in Science and Art.* The Hague: Martinus Nijhoff Pub.
- Edwards, Betty (1986). *Drawing on the Artist Within.* New York: Simon & Schuster Publishing.
- Eliade, Mircea (1963). *Myth and Reality.* New York: Harper & Row.
- Foster, George M. (1962). *Traditional Cultures: and the Impact of Technological Change.* New York: Harper and Brothers Pub.
- Fuchs, Estelle and Robert J. Havighurst (1972). *To Live On This Earth: American Indian Education.* New York: Doubleday.
- Gaffney, K.E., (1992). Multiple assessment for multiple learning styles. *Science Scope*, 15(6), 54-55.
- Gardner, Howard (1982). *Art, Mind and Brain.* New York: Basic Books.
- Gardner, Howard (1983). *Frames of Mind.* New York: Basic Books.
- Garfield, Patricia (1974). *Creative Dreaming.* Simon & Schuster, New York, N.Y.
- Gearing, F. and L. Sangree, ed. (1979). *Toward a Cultural Theory of Education and Schooling.* New York: Mouton Pub.
- Glasser, William (1969). *Schools Without Failure.* New York: Harper & Row.
- Glazer, N. (1981) *Ethnicity and Education: Some Hard Questions.* Phi Delta Kappan, 62, 386-389.
- Gowan, John Curtis (1975). *Trance, Art and Creativity.* Buffalo, N.Y.: Creative Education Association.
- Greene, Rayna (1981). *Culturally-Based Science; The Potential for Traditional People; Science and Folklore.* London: Proceedings of the Centennial Observation of the Folklore Society.
- Hadfield, O.D., Matin, J.V., & Wooden, S. (1992). Mathematics anxiety and learning style of the Navajo middle school student. *School Science and Mathematics*, 92(4), 171-176.
- Hall, Edward T. (1976). *Beyond Culture.* New York: Anchor Press/Doubleday.

- Hall, Edward T. (1984). *The Dance of Life*. New York: AnchorPress/Doubleday.
- Havighurst, Robert James (1955). *American Indian and White Children*. Chicago: University of Chicago Press.
- Havighurst, R.J. (1978). Structural Aspects of Education and Cultural Pluralism. *Education Research Quarterly*, 2 (4): 5-19.
- Hayward, Jeremy W. (1984). *Perceiving Ordinary Magic: Science and Intuitive Wisdom*. Boulder, CO: Shambala Pub. Co.
- Heidlebaugh, Tom and Larry Littlebird (1985). *A Storytelling Workbook for the Disciplines of Making and Having*. Circle Film, Santa Fe, New Mexico.
- Hennessy, S. (1993). *Situated cognition and cognitive apprenticeship: Implications for classroom learning*. Studies in Science Education, 22, 1-41.
- Hill, J.E. (1972). *The Educational Sciences*. Bloomfield Hills, MI: Oakland Community College Press.
- Holbrook, Bruce (1981). *The Stone Monkey*. New York, N.Y.: William Morrow and Company, Inc.
- Honigmann, John J. Ed.(1973). *Handbook of Social and Cultural Anthropology*. Rand McNally College Pub. Co., Chicago.
- Hoover, Kenneth H. (1980). *College Teaching Today: A Handbook for Post-Secondary Instruction*. Boston, MA: Allyn and Bacon, Inc.
- Hughes, J. Donald (1983). *American Indian Education*. El Paso, Texas: Texas Western Press.
- Hultkrantz, Ake (1967). *The Religions of the American Indians*. Translated by Monica Tetterwall 1979, University of California Press, Berkeley, CA.
- Humphreys, B., Johnson, R.T., & Johnson, D.W. (1982). Effects of cooperative, competitive, and individualistic learning on students' achievement in science class. *Journal of Research in Science Teaching,* 19(5), 351-356.
- Hyitfeldt, C. (1986). Traditional culture, perceptual style, and learning: The classroom behavior of Hmong adults. *Adult Education Quarterly*, 36(2), 65-77.
- *Indian Education: A National Tragedy - A National Challenge*. Special Sub-committee on Indian Education, 1969 Report No. 91-501 (Washington: U.S. Government Printing Office, 1969).
- Irvine, J.J. and D.E. York (1995). Learning Styles and Culturally Diverse Students: A Literature Review. (484-497). In Banks, J.E. & C.A. Banks (Eds.). *Handbook Of Research in Multicultural Education*. New York: MacMillan Publishing.

- Jung, Carl G. Ed. (1964). *Man and His Symbols.* London: Aldus Books, Ltd.
- Jenkins, E.W. (1992). *School science education: Towards a reconstruction.* Journal of Curriculum Studies, 24 (3), 229-246.
- Keig, P.F., (1992). *Construction of conceptual understanding in the multicultural science classroom.* Paper presented at "Multicultural Classrooms- A Constructivist Viewpoint," Symposium conducted at San Diego State University, may 20-21, San Diego, CA.
- Kerber, August (1972). *A Cultural Approach to Education.* Dubuque, Iowa:Kendall/Hunt Pub.
- Kessler, C., & Quinn, M.E. (1980). Bilingualism and science problem-solving ability. *Bilingual Education Paper Series*, 4, 1-30.
- Kitano, M.K., (1991). A multicultural perspective on serving the culturally diverse gifted. *Journal for the Education of the Gifted*, 15(1), 4-19
- Kleinfeld, Judy (1972). *Instructional Style and the Intellectual Performance of Indian and Eskimo Students.* University of Alaska at Fairbanks, ERIC Document # Ed. 059-831, 1972.
- Kleinman, Arthur (1980). *Patients and Healers in the Context of Culture.* Los Angeles, California: University of California Press.
- Kleinfield. J.S. (1979) *Eskimo School on the Andreafsky: A Study of Effective Bicultural Education.* New York, N.Y.: Prager Press.
- Knowles, M. (1977). *Modern Practice of Adult Education.* NY: Association Press.
- Kolb, David (1983). *Experiential Learning: Experience as the Source of Learning and Development.* N.J.: Prentice-Hall, Englewood Cliffs, N.J.
- Kovalik, S. (1994). *ITI: The Model-Integrated Thematic Instruction.* Kent: Books for Educators.
- Krasher, S.D. (1981) *Second language acquisition and second language learning.* Oxford: Pergamon.
- Krasher, S.D. (1981). *Second Language Acquisition and Second Language Learning.* Oxford: Pergamon Press.
- Kravagna, Paul Warren (1971). *An Ethonological Approach to Art Education Programming for Navajo and Pueblo Students.* Ed. D. Dissertation: University of New Mexico.
- Kuhn, T.S. (1959). The Essential Tension. *(1959) University of Utah Research Conference on the Identification of Scientific Talent.* Ed. C.W. Taylor. Salt Lake City: University of Utah Press, 1959, pp. 162-174.
- Kuhn, T.S. (1961). The Function of Measurement in Modern Physical Science. *Isis* .52 (1961): 161-90.
- Kuhn, Thomas S. (1962). *The Structure of Scientific Revolutions.* Inter-

national Encyclopedia of Unified Science, Vol. 2, no. 2, Chicago: University of Chicago Press.
• Kuhn, T.S. (1962). The Historical Structure of Scientific Discovery. *Science.* 136 (1962): 760-64.
• The preceding three works are reprinted in: Kuhn, T.S. (1977). *The Essential Tension: Selected Studies in Scientific Tradition and Change.* Chicago: University of Chicago Press.
• Kui Tatk, (Spring 1985). *Newsletter of the Native American Science Education Association.* Washington, D.C.: Vol. 1, No. 1.
• Kunitz, Stephen J. (1981). *The Social Philosophy of John Collier.* New York: University of Rochester.
• Kwagley, A.O. (1995). *A Yupiaq Worldview: A Pathway to Ecology and Spirit.* Prospect Heights: Waveland Press.
• Land, George L. and Vaune Ainsworth-Land (1982). *Forward to Basics.* Buffalo, N.Y.: D.O.K.
• Lawson, Bryan (1980). *How Designers Think.* London: The Architectural Press, Ltd.
• Lore, R.K. (1998). *Art as Developmental Theory: The Spiritual Ecology of Learning and the Influence of Traditional Native American Education.* Ph.d. Dissertation. Albuquerque: University of New Mexico.
• Lomawaima, K.T.(1995). Educating Native Americans. In Banks, J.E. & C.A. Banks (Eds.). *Handbook Of Research in Multicultural Education.* New York: MacMillan Publishing.
• MacIvor, M. (1995). Redefining Science Education for Aboriginal Students. In Bassiste, M. & J. Barman (Eds.). *First Nations Education in Canada: The Circle Unfolds.* Vancouver: University of British Columbia Press.
• Madeja, Stanley S. (1978). *The Arts, Cognition and Basic Skills.* St. Louis: Camrel, Inc.
• Mansfield, Richard S. (1978). *The Psychology of Creativity and Discovery.* Chicago: Nelson - Hall Pub.
• Marriott, Alice and Carol K. Rachlin (1968). *American Indian Mythology.* Harper and Row Pub., New York, N.Y.
• Maruyama, Magorah and Arthur M. Harkins (1978). *Cultures of the Future.* The Hague, Netherlands: Mouton.
• Martinez, D.I., & Martinez, J.V., (1982). *Aspects of American Hispanic and American Indian Involvement in Biomedical Research.* Bethesda, MD: Society for the Advancement of Chicanos and Native Americans in Science.
• McGarry, T.P., (1986). Integrating learning for young children. *Educational Leadership,* 44(3), 64-66.

- McBeth, Sally (August 1984). *The Primer and the Hoe*. New York: American Museum of Natural History. Natural History Magazine 93 (8): 4-12.
- McCarthy, Bernice (1980). *The 4Mat System: Teaching to Learning Styles with Right/Left Mode Techniques*. Barrington, Illinois: Excel, Inc.
- Mintzes, J., J. Wandersee and J. Novak. (1998) *Teaching Science for Understanding*. San Diego: Academic Press.
- Meriam, Lewis (1928). *Problems of Indian Administration*. Baltimore, Maryland: Johns Hopkins University.
- Miller, Arthur I. (1984). *Imagery in Scientific Thought creating 20th Century Physics*. Boston, MA: Birkhauser Publishing Co.
- More, A.J. (1990). *Learning Styles of Native Americans and Asians*. (Report N0. RC-018-091). Vancouver, CA: University of British Columbia. (ERIC Document Reproduction Service No. ED 330 535)
- Morey, Sylvester M. and Olivia L. Gilliam ed. (1974). *Respect for Life: The Traditional Upbringing of American Indian Children*. Freeport, N.Y.: Waldorf Press - Castlereagh Press, Inc.
- Myerhoff, Barbara G. (1974) *Peyote Hunt: The Sacred Journey of the Huichol Indians*. Ithaca N Y Cornell University Press
- Myers, Gail E. and Michele T. Myers (1973). *The Dynamics of Human Communication*. New York: McGraw-Hill, Inc.
- Naroll, Raoul and Frada Naroll, ed. (1973). *Main Currents in Cultural Anthropology*. New York: Appleton - Century Crofts.
- New, Lloyd (1968). *Cultural Difference as the Basis for Creative Education: Native American Arts*. No. 1. Washington, D.C.: U.S. Department of the Interior, Indian Arts and Crafts Board, Government Printing Office.
- Ogbu, J.U., (1992). Understanding cultural diversity and learning. *Educational Researcher*, 21(8), 5-14.
- Okebbukola, P.A. (1986). The influence of preferred learning styles on cooperative learning in science. *Science Education*, 70(5), 509-517.
- Ornstein-Galicia, J.L., & Penfield, J. (1981). A problem-solving model for integrating science and language in bilingual/bicultural education. *Bilingual Education Paper Series*, 5, 1-
- Orr, David W. (1994). *Earth in Mind: On Education, Environment, and the Human Prospect*. Washington, D.C.: Island Press.
- Ovando, C.J. & Collier, V. (1985). *Bilingual and ESL classrooms: Teaching in multicultural contexts*. New York: McGraw-Hill.
- Ovando, Carlos J. (1992). Science. In Reyner, J. (Ed.) *Teaching American Indian Students*. Norman: University of Oklahoma Press.
- Peat, F. David (1996). *Lighting the Seventh Fire: The Spiritual Ways, Healing, and Science of the Native American*. New York: A Birch Lane Press.

- Pepper, Floy C. (1985). *Understanding Indian Students: Behavioral Learning Styles*. Portland Oregon: Research Development for Indian Education, Northwest Regional Educational Laboratory, Pub.
- Pettitt, George A. (1946). *A Primitive Education in North America*. University of California Publications in Archeology and Ethology, XLIII.
- Phelan, P., Davidson, A. & Cao, H. (1991). Students' multiple worlds: Negotiating the boundaries of family, peer, and school cultures. *Anthropology and Education Quarterly*, 22(3), 224-250.
- Polieoff, Stephen P. (1985). In Search of the Elusive Aha!. *New Age Journal*, March 1985: 43-39. Brighton: Rising Star Assoc.
- Pomeroy, D. (1992). *Science Across Cultures*. Chicago: University of Chicago Press.
- Pitman, M.A., (1989). *Culture acquisition: A holistic approach to human learning*. New York: Praeger.
- Read, Herbert (1945). *Education Through Art*. New York: Pantheon Books.
- Reyner, J. (Ed.) (1992). *Teaching American Indian Students*. Norman: University of Oklahoma Press.
- Rico, Gabriele L. (1983). *Writing the Natural Way*. Los Angeles: J.P. Tarcher, Inc.
- Rhodes, R.W. (1988). Holistic teaching/learning for Native- American students. *Journal of American-Indian Education*, 27(2), 21-29.
- Roberts, Joan I. and Sherril Akinsanya (1976). *Educational Patterns and Cultural Configurations: The Anthropology of Education*. New York: D. McKay Co.
- Roberts, S. and S.A. Akinsanya, ed. (1977). *Schooling in the Cultural context: Anthropological Studies of Education*. N.Y.: D. McKay Pub.
- Rodriguez, I., & Bethel, L.J. (1983). An inquiry approach to science/language teaching. *Journal of Research in Science Teaching*, 20 (2), 291-296.
- Ronan, C.A. (1982). *Science: Its history and development among the world's cultures*. New York: Hamlyn Publishing.
- Rothenberg, Albert (1982). *The Emerging Goddess: The Creative Process in Art, Science and Other Fields*. New York: Nelson-Hall Pub.
- Stahl, R.J. (1992, Mar.). *Using the information-constructivist (IC) perspective to guide curricular and instructional decisions toward attaining desired student outcomes of science education*. Paper presented to the Annual Meeting of the National Research in Science Teaching, Boston, MA.
- Strum, F. & T. Purley. (1985). Pueblo valuing in transition. *Pueblo Cultural Center Newsletter, Vol. 10., No. 12.*.

- Sanders, Donald A. (1986). *Teaching Creativity Through Metaphors: An Integrated Brain Approach.* New York: Longman Pub.
- Saxe. G.B. and Posner (1983). The Development of Numerical Cognition: Cross-Cultural Perspectives. (291-317). *The Development of Mathematical Thinking.* Ginsburg, H.P. (Ed.) New York: Academic Press.
- Scheffler, Israel (1967). *Science and Subjectivity.* Indianapolis: Bobbs-Merrill Co.
- Scherer, Marge (January 1985) How Many Ways is a Child Intelligent? :An interview with Howard Gardner. *Instructor Magazine* 94 (5): 32-35. New York: Instructor Publication, Inc.
- Scott. James Michael (1975). *The Creative Process in Scientific and Artistic Problem Solving.* MFA Dissertation. University of New Mexico.
- Shimkin, D.B., S. Tax and J.W. Morrison ed. (1978). *Anthropology for the Future.* R.R. #4, Dept. of Anthropology, University of Illinois, Urbana, Illinois.
- Snow, Albert J. (1972). Ethnoscience in American Indian Education. *The Science Teacher Magazine*, vol. 39, No. 7, October 1972: 30-32.
- Snow, Albert (1974). *American Indian Ethnoscience: A Study of Its Affects on Student Achievement.* Ed. D. Dissertation, University of Maryland.
- Smith, G.A. and D.R. Willaims (Eds.). (1999). *Ecological Education in Action.* Albany: State University of New York Press.
- Spindler, George G. (1963). *Education and Culture.* New York: Holt, Rinehart and Winston.
- Spindler, George, ed. (1982). *Doing the Ethnography of Schooling.* New York: Holt, Rinehart and Winston.
- Spradley, James, Ed. (1972). *Culture and Cognition.* New York: Chandler Publishing Co.
- Sprague, Arthur William (1972). *Additional Changes in Secondary School Students as a Result of Studying Ethnohistory of Kiowa Indians.* Ph.D. Dissertation, Ohio State University.
- Steiner, Vera John (1986). *Notebooks of the Mind.* Albuquerque: University of New Mexico Press.
- Szasz, Margaret Connell (1977). *Education and the American Indian.* Albuquerque: University of New Mexico Press.
- Tippens, D.J., & Dana,N.F. (1992). Culturally relevant alternative assessment. *Science Scope*, 15(6), 50-53.
- Thomas, Lewis (1974). *The Lives of a Cell.* New York: Viking Press.
- Torrance, Paul E. (1963). *Education and the Creative Potential.* Minneapolis: University of Minnesota Press.
- Torrance, Paul (1980). *Your Style of Learning and Thinking: Learning Styles Inventory.* Excel, Inc., Barrington, IL.

- Tuan, Yi-Fu (1974). *Topophllia: A Study of Environmental Perceptions, Attitudes, and Values.* Englewood Cliffs: Prentice-Hall, Inc.
- Turpin, Thomas Jerry (1975). *The Cheyenne Worldview as Reflected in the Stories of Their Culture Heroes, Erect Horns and Sweet Medicine.* Ph.D. Dissertation: University of Southern California.
- UNESCO (1978) *Transfer of Knowledge Research News Letter.* UNESCO Division for the Study of Development, New York.
- Van Peursen, C.A. (1981). *Creativity as Learning Process in the Concept of Creativity in Science and Art.* The Hague: Martinus Nijhoff Pub.
- Vitale, Barbara Meister (1982). *Unicorns are Real : A Right Brained Approach to Learning.* California: Salmar Press.
- Von Bertalanffy, Ludwig (1968). *General Systems Theory.* New York: George Braziller, Inc.
- Vygotsky, L.S. (1978). *Mind and Society.* Cambridge: Harvard University Press.
- Valle, R. (1986). Cross-cultural competence in minority communities: A curriculum implementation strategy. In M.R. Miranda & H.H.L. Kitano (Eds)., *Mental health research and practice in minority communities: Development of culturally sensitive training programs.* (pp. 29-49). Rockville, MD: National Institute of Mental Health.
- Van Otten, G.A., & Tsutsui, S. (1983). Geocentrism and Indian education. *Journal of American-Indian Education,* 22(2), 23-27.
- Weatherford, J. (1988). *Indian Givers: How the Indians of America Transformed the World.* New York: Fawcett Columbine.
- Wax, Murray (1971). *Indian Americans: Unity and Diversity.* Englewood Cliffs: Prentice-Hall.
- Weinrich, U. (1953). *Languages in Contact: Findings and Problems.* New York: Linguistic Circle of New York.
- Whiteman, Henrietta (1985). *Historical Review of Indian Education.* Ninth Inter-American Indian Congress, October 1985, Santa Fe, N.M.
- Whorf, Benjamin Lee (1956). *Language, Thought and Reality._*Cambridge: M.I.T. Technology Press.
- Wilber, Ken (1982). *The Holographic Paradigm and Other Paradoxes.* Boulder: Shambala Press.
- Zak, Nancy C. (1984). *Sacred and Legendary Women of Native America. Part I & II,* (Newsletter).Volume 1, No. 1 & 2, November 15, 1984 and March 15, 1985, pp. 12-14 and 8-11.
- Zais, Robert (1976). *Curriculum Principles and Foundations.* New York, N.Y.: Harper & Row.